The Complete Handbook of Amplifiers, Oscillators and Multivibrators

by Joseph J. Carr

TAB BOOKS Inc.
BLUE RIDGE SUMMIT, PA. 17214

FIRST EDITION

THIRD PRINTING

Printed in the United States of America

Reproduction or publication of the content in any manner, without express permission of the publisher, is prohibited. No liability is assumed with respect to the use of the information herein.

Copyright © 1981 by TAB BOOKS Inc.

Library of Congress Cataloging in Publication Data

Carr, Joseph J
    The complete handbook of amplifiers, oscillators,
and multivibrators.

    "Tab book # 1230."
    Includes index.
    1. Amplifiers (Electronics)   2. Oscillators,
Electric.   3. Multivibrators,   I. Title.
TK7871.2.C34   621.3815'3   80-28375
ISBN 0-8306-9653-9
ISBN 0-8306-1230-0 (pbk.)

Contents

Introduction

This book is a handbook on the practical design of, and theory behind, certain basic electronics building blocks: amplifiers, oscillators and multivibrators. These circuits form the basis for most electronic equipment and projects that might be undertaken. Understanding these circuits, then, is critical to understanding electronic instruments and equipment. Whether your area be amateur radio, professional communications, engineering, hobby electronics, or class work as a student, you should find this book useful. Students in engineering and electronics technology courses will find the material particularly useful for aiding in the design of the senior project (required by many schools).

Amplifiers are familiar to most persons who would buy a book like this. We will cover the basic theory behind most forms of common amplifier devices. We will examine bipolar transistors, junction field-effect transistors, metal oxide field effect transistors (MOSFETs), operational amplifiers, linear ICs and so forth. We will discuss some basic theory common to all forms of amplifiers. There are three basic classification systems in use: *conduction angle* (class-A, class-B, etc), *common electrode* (common emitter, common gate, etc.), and *transfer function* (voltage, current, transresistance and transconductance amplifiers). These different classes will be discussed in terms of one transistor or another such as bipolar, JFET or MOSFET, but the same concepts are easily transfered to the other types of active devices by swapping equivalent terminals and labels. For example, common emitter in the bipolar transistor is analogous to the common source of the JFET/MOSFET devices.

Oscillators and multivibrators are sometimes regarded as the same thing. But in this book, as in most areas of electronic technology, it is common practice to label as *oscillators* those circuits which produce a sine wave output. We find LC and RC versions operating at frequencies from less than 1 Hz to dozens of megahertz. The multivibrator, on the other hand, is a nonlinear waveform generator. These circuits produce square waves, triangle waves, pulses and an assortment of other nonsinusoidal waveshapes. The multivibrator is usually an RC circuit, coupled with an active device. Only a few LC circuits are still used in this area.

You will learn little in the way of heavy theory, but you will cover the operation of the devices and circuits from a descriptive point of view, with rule-of-thumb design techniques. We have not avoided the use of arithmetic where appropriate, but we also refuse to lean on the math when a little prose will go a long way toward explaining the operation of a circuit that is also the subject of an equation. It is hoped that this approach will serve many masters.

Joseph J. Carr

Other TAB books by the author:

Chapter 1
Basic Semiconductor
Theory—The pn Junction

The transistor has all but totally eclipsed the vacuum tube. For several decades, the tube reigned supreme, and exclusively. But in the late 1940s physicists at Bell Laboratories invented the transistor. Fortunately for the electronics industry as a whole, Bell Labs wisely decided to license the device to other manufacturers and teach them the technology required to produce these little devices. Today it is difficult to find any new products that still use vacuum tubes. In the radio transmitter industry—the last hold out—the power level of transistor transmitter final amplifiers has been increasing to the point where it is conceivable to make a final amplifier in the 1000-watt class with transistor devices. This is the only area where vacuum tubes are still the "in" thing. Transistor manufacturers have yet to find a way to make multikilowatt solid-state final amplifiers at anything like a competitive price with vacuum tubes.

The material in this chapter is essentially a review for most readers. Although a little review never hurt anybody, it is probable that more knowledgeable readers will want to go on to one of the other chapters. The theory here is abbreviated somewhat for the sake of saving both space and the reader's time. Those wishing to obtain a more in-depth viewpoint are referred to Andy Veronis' excellent book, TAB book No. 717, *Transistor Theory for Technicians & Engineers*. Veronis goes more deeply into the physics of semiconductors as it relates to the design and construction of transistor devices.

The transistor was not actually *invented* until the late 1940s, but it had been postulated as theoretically possible for at least a decade prior to its invention. The metallurgy of the time, however, was not sufficiently advanced to make the purity of materials needed in transistor construction. One popular story that floats around the industry tells us that the function field-effect transistor (JFET) was "invented" on paper as a physicist's *thought experiment* in the late 1930s. But, again, metallurgy was the problem. It has been said that the JFET has certain properties that are superior to the bipolar transistor, and would have been the standard transistor if the metallurgists could have produced the device. This would have given our present electronics industry a decade lead over its present state of development!

The transistor is a descendant of the simple *pn junction diode*. These diodes consist of two sections of semiconductor material, one being given negative charge carriers (*electrons*) and the other positive charge carriers (*holes*). The transistor is likened to a pair of pn junction diodes connected back-to-back, but more of that later. The pn junction diode is found in nature. Certain minerals, such as the lead oxide called *galena*, will operate as a pn junction diode. The early crystal radio sets, widely used in the years before vacuum-tube amplifiers made radio a *real* practicality, used chunks of galena to demodulate the radio signal. Artificial pn junctions became available in the early days of radio. These were made of copper oxide, or *selenium*. But it took the metallurgical advances under the pressure of World War II to make manufacturing pn junction diodes which operated into the radio spectrum an economic reality. The 1N34 and 1N60 germanium diodes and the 1N21 silicon diode were developed through this effort. The 1N21 was a VHF microwave device that was used extensively in radar receivers as the mixer diode. With the technology of World War II available, the physicists at Bell Labs were finally able to make the first transistor.

The transistor was a slow starter in the electronics world. When it was born, there were still severe manufacturing problems to overcome, and yield rates (the percentage of a batch that actually worked up to specifications) were low—quite low, in fact. The *1955 Allied Electronics Catalog* featured the CK722 and 2N107 transistors for about $16. And remember the dollar was worth more then! These leaky germanium transistors operated only at audio frequencies and would easily burn out if abused. And it took a lot less to abuse those devices than it does with modern devices.

We were warned to use heatsink clips on the leads of the transistor when soldering, lest we melt the semiconductor material inside. Only a few rf-range transistors were available, and these were expensive. The first all-transistor portable radios were seen in 1954 to 1955, with a Motorola car radio model in 1956. It wasn't until 1962, however, that a major American automobile manufacturer introduced the first production OEM all-transistor car radio (Delco Electronics of General Motors), followed by the rest of the makers the following year. The *all-American-five* home radio, a standard vacuum-tube design (which used the same tube line-up, no matter who made it or what style case was used), was still made until the mid-60s. The problem was that these radios used no power transformer and operated by directly rectifying the 115 VAC power line. It took the introduction of high-voltage audio power transistors to make it possible to manufacture a low-priced, AC/DC table radio.

ELEMENTARY SEMICONDUCTOR THEORY

We must start any study of semiconductors with the basic model of the atom. We use a model, now known to be over-simplified, in which a nucleus of positively charge and neutral particles is surrounded by a cloud of electrons orbiting in *shells*. The electrons carry a negative charge of the same magnitude but opposite polarity of the charge on the positively charged particles in the nucleus.

The elementary nucleus consists of positively charged particles called *protons*, and electrically neutral particles called *neutrons*. The protons and neutrons have almost the same mass, but they are approximately 1850 times heavier than the orbiting electrons. The amount of charge, however, is the same, although it is opposite.

The cloud of orbiting electrons contains as many electrons as there are protons in the nucleus. This arrangement permits *electroneutrality* (nature just loves order and balance), which means that the positive charges in the nucleus are exactly balanced; therefore, they are effectively cancelled, by the negative charges of the electrons.

The electrons orbit in shells, which is to say fixed distances from the nucleus. These shells are the only distances at which the electrons are allowed to exist, and they represent the energy level of the electrons in the shell. When an electron takes on sufficient energy, it will jump to the higher shell—the one farther from the nucleus—that corresponds to that level. When an electron falls from an outer shell to an inner shell, it loses energy. Because we must maintain the law of conservation of energy, we find that the "lost" energy is actually given off in the form of light, infrared, or some other form of energy.

It appears that each shell is ideally filled with an exact number of electrons. When this number is reached, that shell is stable and any additional electrons begin to fill a shell that is farther from the nucleus. Chemical reactions, and the flow of electricity, depend mostly on the last shell in any given element. When the outermost shell is filled with its ideal number of electrons, a large amount of energy is needed to strip electrons loose for participation in the reaction known as current flow. These electrons are called *valence electrons*. How many are required to make the shell stable? For the shell closest to the nucleus, the stable configuration is two electrons. Hydrogen has just one electron in this shell, while helium fills the shell with two electrons. The hydrogen atom is chemically active because its outer—indeed, only—shell is filled with less than the ideal number of electrons. Helium, on the other hand, has two electrons in this shell, so it is ideally filled for stability. Helium is not chemically active. The second shell is ideally filled with eight electrons, while the third shell is filled with 18 electrons and so forth.

An ideally filled shell does not want to give up electrons. It takes a large amount of energy to make such a shell deliver electrons. If the outermost shell of an atom is *not* ideally filled, however, the outer electrons are not tightly bound. Consequently, only a small amount of energy is necessary to strip electrons from that atom. Consider carbon, which contains six electrons. Two of the electrons will completely fill the innermost shell, leaving four electrons for the second shell. Since this shell "wants" to have eight electrons, it will be loosely bound and will give up electrons easily. These four outer electrons give carbon its ability to permit the flow of electrical current and to participate in chemical reactions.

In electronics we recognize three classes of materials: *insulators, conductors* and *semiconductors*. As you might have guessed, insulators are those materials that have atoms with ideally filled outer shells. They require a large electrical potential field to strip away *free electrons* for current flow. Consequently, they will allow only a tiny leakage current to pass through them. Insulators are available for which *picoamperes* is too large a unit to describe the level of current flow even under potential differences of kilovolts! The outer shells of atoms in materials called conductors are loosely bound, which means they do not contain the ideal number required for filling. These atoms can have the outer electrons stripped away relatively easily, with only a small amount of energy needed. A little heat in some cases or a low voltage will easily remove an electron or two from these atoms.

Semiconductors are elements which are neither good conductors, nor good insulators. Examples of semiconductors include germanium (Ge) and silicon (Si). Both of these elements are used in the manufacture of diodes, transistors and other semiconductor products. Germanium has an atomic number of 32, meaning that it contains 32 protons balanced by 32 orbiting electrons. There are two electrons in the innermost shell, eight in the second shell, 18 in the third shell and only four in the fourth, or outermost, shell.

Silicon has an atomic number of 14, meaning that its 14 protons in the nucleus are balanced by 14 electrons in the orbiting cloud. There are two electrons in the innermost shell and eight in the second shell. Both of these shells are ideally filled. The outer shell takes the remaining four electrons, which makes it less than ideally filled.

Both silicon and germanium have only four electrons in the outer shell; therefore, both have four valence electrons. As such, these elements are said to be *tetravalent*. Shells can be made stable with eight electrons. Tetravalent atoms will try to mimic this stable configuration by forming *covalent bonds* between adjacent atoms in a crystalline structure. Figure 1-1 shows how this is done. Note that the actual situation is three dimensional, but to keep this discussion simple, we are going to demonstrate only a two-dimensional plane. Each atom has four valence electrons, which are each shared with one electron of an adjacent atom. The result is a simulation of the stable octet. Note that each bond contains two electrons, one contributed by each of the atoms forming the bond.

The bonding makes the element more stable, so there are few free electrons to participate in a current flow. There are a few free electrons floating around but not nearly as many as in conductors. Consider the situation if certain types of impurities are added to the pure semiconductor material, though. Suppose that the tetravalent semiconductor material (either germanium or silicon) is *doped* with a *pentavalent* (five valence electrons) impurity (phosphorous, arsenic, or antimony). The rule requiring electroneutrality is not violated because each pentavalent atom is, in itself, electrically neutral; it contains as many electrons as there are protons. Figure 1-2 shows how a pentavalent atom forms covalent bonds in a tetravalent crystal (only one atom is shown with its four covalent atoms for the sake of simplicity). Because five valence electrons are in a pentavalent material, there is one *excess* electron for each pentavalent atom. This is because only four of the electrons in this atom can find "mates" in the tetravalent world it sees. The electrons that do not form covalent bonds with nearby tetravalent atoms are available are charge carriers in the flow of electrical current. A material doped with a pentavalent impurity is an *n-type* semiconductor, and its *charge carriers* are *electrons*.

P-type semiconductor material is opposite the n-type: its charge carriers *appear* to be positively charged particles. The p-type semiconductor is also doped with an impurity, but in this case a *trivalent* material is used in order to create an electron deficiency. The configuration of p-type semiconductor material is shown in Fig. 1-3. Each trivalent atom contains three valence electrons that will form covalent bonds with the valence electrons of the tetravalent atoms; however, only three tetravalent atoms can form such bonds with the trivalent atoms. This leaves a *hole* in the crystal structure that seeks to be filled to become stable. The simplest definition of a *hole* is it is "a place where an electron should be, but isn't."

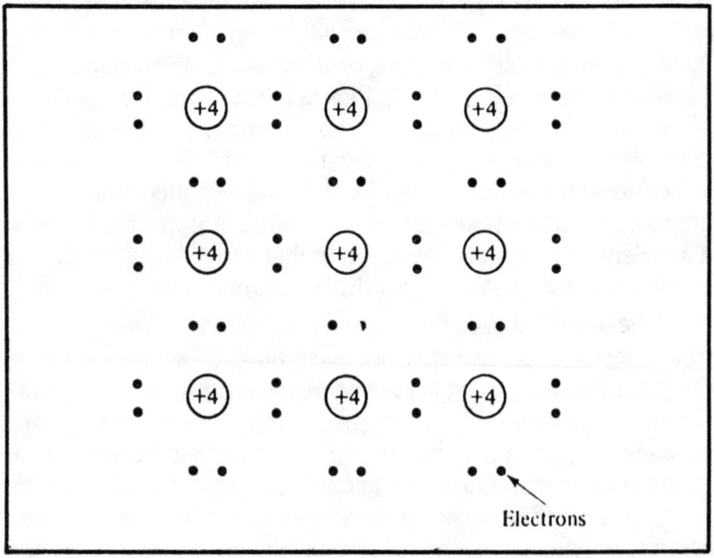

Fig. 1-1. Structure of the semiconductor crystal.

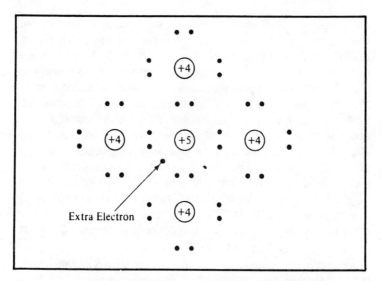

Fig. 1-2. Addition of pentavalent atom creates an excess electron (N-type).

The concept of holes usually gives students difficulty. It really should not, however, because it is essentially a very easy concept. Part of the problem is that many electronics textbooks refer to the holes exclusively in terms of positive charge carriers that *flow*. Holes don't flow; they only *appear* to flow. The electrons actually do the flowing. We treat the p-type semiconductor material *as if*

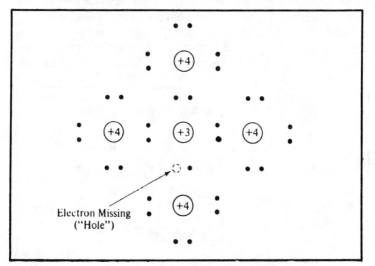

Fig. 1-3. Addition of a trivalent atom creates an electron deficiency (P-type).

holes—positively charged particles—were actually flowing, but this is merely a convention of convenience. The situation is shown in Fig. 1-4. An electron can fill a hole in one of the bonds that are deficient, but only at the expense of creating a hole somewhere else. This makes it appear that the hole "flowed" from one atom to the other, but it was actually an electron that did all the moving. Consider Fig. 1-4. At the instant that voltage is applied across the doped semiconductor material (p-type) there is a hole located in atom A. The voltage field acts as a force that breaks loose an electron from atom B. This electron will drift under influence of the electric field until it is captured by atom A to fill the hole. We have filled a hole (obliterating it) at atom A, but only at the expense of creating a hole at atom B. The hole didn't move. Suppose that, a short time later, an electron is forced loose from atom C, and it drifts through the material until it fills the hole at atom B. This action creates a hole at atom C but fills one at atom B. Again, there is an *apparent* movement of the hole from atom B to atom C. This action continues for some length of time, so that we see an apparent hole drift over the path to atom F. Once again, the actual moving was done by a succession of several electrons. To an external observer looking for positive charges, though, the appearance is that some sort of positive charge with the mass of an electron has moved. But this is in *appearance* only. We sometimes call any charges merely positive or negative *carriers*. In p-type material, the carriers are the holes, while in n-type material the carriers are electrons. Both of these carriers have a net charge of one unit, although they have opposite polarity. Also, p-type material contains some negative charge carriers, while n-type material contains some positive carriers; however, these materials contain

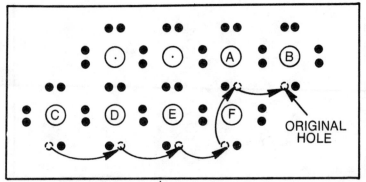

Fig. 1-4. "Hole flow" is actually electron flow viewed in reverse.

predominantly more of their respective charges. The charge flow in n-type material is from negative to positive terminals of the power supply, while the hole flow is just the opposite.

THE PN JUNCTION

A pn junction consists of a section of p-type material interfaced with a section of n-type material. The basic model of an unbiased pn junction is shown in Fig. 1-5A. In the earliest, crude junction diodes, these sections were made separately and then joined together. Today, the practice is to make the pn junction from a monolithic piece of germanium or silicon, with the pentavalent and trivalent impurities diffused into the n-type and p-type sections, respectively. The germanium diode was the first to be commercially available outside of the radar business, but it has been largely eclipsed by the more thermally stable silicon diode.

The model for a *reverse-biased* pn junction is shown in Fig. 1-5B. In this case, the positive terminal of the battery is connected to the n-type material, and the negative terminal of the battery is connected to the p-type material. Recall that like charges repel each other and unlike charges attract each other. In the reverse bias situation, the negative charges of the n-type material are attracted to the positive terminal of the battery, and the positive charges of the p-type material are attracted to the negative terminal of the battery.

There is a zone near the junction in which all of the carriers have been removed, called a *depletion zone*. Very few carriers of either polarity exist inside the depletion zone. The depletion zone is deficient in carriers, so it has a very high electrical resistance. Little or no current will flow across the pn junction under this condition. Only a very small leakage current exists (leakage is also somes called *reverse current*).

The model for a forward biased pn junction is shown in Fig. 1-5C. In this case, the polarity of the battery is reversed from the previous case: The positive terminal of the battery is connected to the p-type material, and the negative terminal is connected to the n-type material. Following the like-charges-repel rule, the positive carriers in the p-type are driven away from the battery and toward the junction. Similarly, the negative carriers are driven away from their battery terminal toward the junction. The depletion zone width reduces to zero, and charges can combine across the junction. This causes a current flow across the junction.

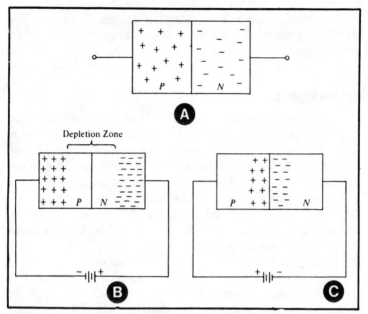

Fig. 1-5. (A) Unbiased pn junction, (B) reverse biased pn junction and (C) forward biased pn junction.

If a pn junction is reverse-biased, no current will flow. If a pn junction is forward-biased, current will flow. The diode current is given by:

$$I = I_s [e^{(V/nV_t)} - 1] \qquad \textbf{Equation 1-1}$$

where I is the diode current, I_s is the reverse saturation current (leakage), e is the base of the natural logarithms, V is the voltage across the function, n is a constant (approximately) 1 for Ge and 2 for Si), V_t is the volt equivalent of temperature, and is defined as $V_t = KT/q$, where K is Boltzmann's constant (1.38×10^{-23} joules/°K), T is the temperature in degrees Kelvin, and q is the electronic charge (1.6×10^{-19} coulombs). The value of V_t evaluates to $V_t = T/11600$. If we assume room temperature to be 300°K, $V_t = 0.026$ volts, or 26 millivolts.

☐ **Example:**

Find the current flow in a silicon pn junction diode at room temperature if the applied voltage is 1.6 volts DC. Assume a reverse saturation current of 10^{-14} amperes.

$$I = I_s (e^{V/nV_t} - 1)$$
$$I = (10^{-14})(e^{(1.5/(2)(0.026))} - 1)$$
$$I = (10^{-14})(e^{(28.8)} - 1)$$
$$I = (10^{-14})(3.37 \times 10^{12}) = 0.0337 \text{ amperes} = 33.7 \text{ mA}$$

Figure 1-6 shows the current-vs-voltage characteristic for a pn junction diode. There are several regions of interest: *reverse bias, forward bias, saturation, breakdown* and *puncture*. The reverse bias region portrays the situation when the voltage applied to the diode will cause no current flow, as in Fig. 1-5. The only current flow will be the reverse saturation current, or leakage current, I_s. At some voltage V_z, however, the diode will break down in the reverse direction and a reverse current will flow. This voltage is called the *peak inverse voltage* (PIV) or *peak reverse voltage* (PRV). In a certain class of voltage regulator diodes called zener diodes, the breakdown is controlled, and the transition knee is sharper.

The forward current in a pn junction is not linear at low forward bias levels. The current is a nonlinear function of the applied voltage at potential levels between zero and some positive *junction voltage*, V_j. At voltages greater than V_j, however, the transfer curve straightens out and becomes linear. At some high

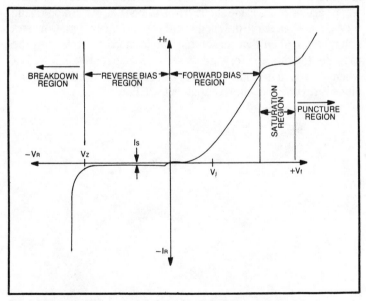

Fig. 1-6. I-versus-V characteristic for pn junction diode.

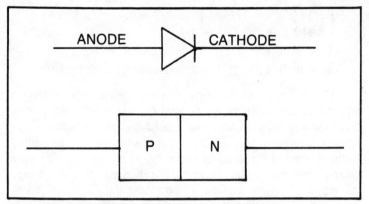

Fig. 1-7. Symbol for the pn junction diode and block diagram.

value of forward bias no additional current will flow for increases in applied voltage. This is the saturation region. Finally, at some high forward voltage, the junction will break down (*punch through region* or *puncture region*).

Figure 1-7 shows the symbol commonly used for pn junction diodes used in electronics. The arrow indicates the direction of convention current flow, which is exactly the opposite of electron flow. This difference, incidentally, has confused electronics students for decades, ever since electron flow became important when vacuum tubes came about. The cathode end of the diode is the n-type material, while the anode end is the p-type material. When a voltage is applied that makes the anode more positive than the cathode, the diode is forward biased and a current will flow. But when the anode is made negative with respect to the cathode, the diode is reverse biased and no current will flow.

Chapter 2
Basic Transistor Theory

The transistor was invented in the late 1940s. By now it has surpassed the vacuum tube in all applications other than high-power transmitters. Even in transmitters, however, solid-state power amplifiers to the kilowatt level are relatively easy to obtain. The transistor is generally easier to apply and more efficient in several ways than tubes. Transistors are physically smaller and far more adaptable to printed circuit mounting than are tubes. They require no filament power, so they are inherently more efficient that tubes that do the same job—and *do* require filament power. The operating voltages for transistors are quite low. Transistors are usually happy with the type of voltages available from simple batteries, making them ideal for application in mobile or portable equipment.

The transistor is sometimes incorrectly modeled as a pair of pn junction diodes connected back-to-back. Figures 2-1 and 2-2 show the two basic types of bipolar transistor in model form. Both transistors can be modeled as having three sections. In both cases, a section of one type semiconductor material is sandwiched between two sections of the opposite material. The npn transistor is shown in Fig. 2-1, along with its circuit symbol. Note that the arrow on the emitter terminal is *pointing out* for the npn transistor. In the npn transistor, the center section (base) is made of p-type semiconductor material, and is sandwiched between n-type end sections (emitter and collector, respectively).

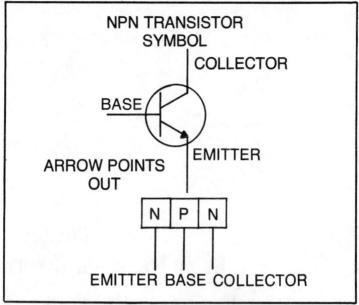

Fig. 2-1. The npn transistor symbol and block diagram.

The pnp transistor is exactly the opposite of the npn. In this case, shown in Fig. 2-2, the base section is made of n-type material and is sandwiched between p-type emitter and collector sections. The symbol for a pnp transistor is also shown in Fig. 2-2. A little memory aid for remembering which symbol is to be used is *pnp* = *P*oints i*N*.

Figure 2-3 shows an npn transistor biased for normal amplifier operation. The emitter-base junction is forward biased, and the collector is made positive with respect to the emitter. Note that biasing for a pnp equivalent amplifier is exactly the same, except that the battery polarities are reversed.

The circuit action in a transistor is a little more complex than our more simple descriptions. In general, though, we can claim that the charge carriers attracted into the base region from the emitter pass through to the collector. Only a small percentage of the current passes out of the base terminal of the transistor to the external circuit; on the order of 95 to 99 percent of the emitter current flows also in the collector circuit. The remaining 1 to 5 percent passes out of the transistor through the base circuit.

Amplification

Amplification is the control of a larger voltage or current by a

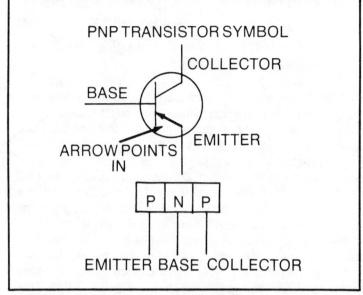

Fig. 2-2. The pnp transistor symbol and block diagram.

smaller voltage or current. In the case of the transistor circuits presented here, a small *base* current is capable of controlling a larger collector-emitter current.

The voltage appearing across the base-emitter junction is given by an expression that is derived from Equation 1-1 in the previous chapter. In this case, however, the expression for current is solved to give V, which this time is given as V_{be}:

$$V_{be} = \frac{KT}{q} \quad \ln \left[\frac{I_c}{I_s} \right] \qquad \textbf{Equation 2-1}$$

where V_{be} is the base-emitter voltage, K is Boltzmann's constant $(1.38 \times 10^{-23}$ J/C, T is the temperature in degrees Kelvin, q is the electronic charge $(1.6 \times 10^{3-19}$ coulombs), I_c is the transistor collector current, I_s is the reverse saturation current (approximately 10^{-10} amperes), and l_n denotes the use of natural (base-e) logarithms.

□ **Example:**

A transistor with a reverse saturation current of 10^{-12} A is operated at a temperature of 300° K. Find the base-emitter voltage when the collector current is 2 mA.

$$V_{be} = (KT/q)(\ln(I_c/I_s)$$
$$= (\ (1.38 \times 10^{-23} \text{ joules/coulomb}) \ (300° \text{ K})$$
$$/1.6 \times 10^{-19} \text{ coulombs})) / (\text{Ln}((2 \times 10^{-3}A)/(10^{-12}))$$
$$= (0.026) \ (\ln(21.4) = 0.56 \text{ volts}$$

The ordinary base-emitter voltage for a properly forward-biased transistor will fall within a narrow range that is dependent upon the semiconductor material used. Germanium transistors, for example, usually exhibit a junction potential of 0.2 to 0.3 volts, while silicon transistors exhibit a potential of 0.6 to 0.7 volts.

What happens when a signal source is applied to the input of the transistor amplifier? Figure 2-4 shows the circuit with a AC signal source connected in series with the collector-emitter circuit. As in all of our discussions, unless stated otherwise, the signal source is a sinusoidal waveform. It will add to or subtract from the ordinary base bias current. On positive excursions of the input sine wave, the collector-emitter current will increase because the current produced by the signal source adds to the base current. Similarly, on negative excursions of the input signal, the collector-emitter current will decrease because the input signal subtracts from the bias current. We can make the claim that the transistor is a basic current amplifier because I_b is always less than I_c; in some cases it is *very* much less.

Let's now clear up a problem before it becomes firmly rooted in somebody's mind. In practice, the transistor can also be used as a voltage amplifier as well as a current amplifier, but many students have that current amplifier idea so firmly embedded into their minds that they seem unable to grasp the idea of voltage

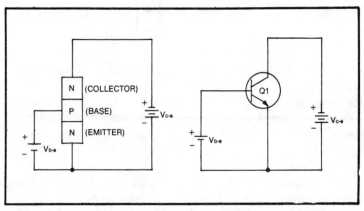

Fig. 2-3. The npn transistor showing biasing scheme for proper operation.

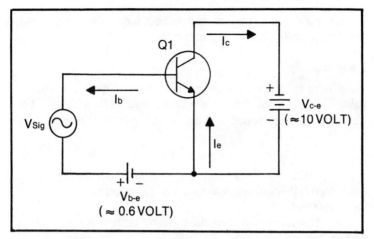

Fig. 2-4. Forward biased transistor with signal superimposed on base current.

amplification. For years, some people wandered about the landscape muddering "tubes are voltage amplifiers, and transistors are current amplifiers." Set that little myth to rest right now.

First, recall *Ohm's law*, which relates voltage to the product of the current and resistance in circuit (E x IR). Consider the circuit shown in Fig. 2-5. Resistor R1 in Fig. 2-5 is connected in series with the collector of the transistor and the power supply. The power supply potential is V_{cc}, while the collector-to-emitter voltage is V_c. The difference between V_c and V_{cc} is the voltage drop across collector load resistor R1, which is I_cR_1. The output voltage is the collector voltage, V_c x $V_{cc} - (I_cR_1)$. Again, assume that the signal voltage is a sinudoid. As V_{sig} increases, then so does collector current I_c. The collector voltage must then drop because collector current increases the voltage drop across the collector load resistor (I_cR_1), leaving less of the supply voltage for V_c. Similarly, when the signal voltage decreases, the collector current decreases, which causes the collector voltage to rise. In this case, the voltage drop across the collector load resistor is less as a result of the lower collector current, and there is more of the battery voltage available to make an ouput signal. The input signal voltage is less than the swing in the DC collector ouput voltage, so the transistor is operating as a voltage amplifier.

Thus far, we have turned our attention to the npn transistor, but the same principles can be applied also to the pnp device. It is merely necessary to reverse both the polarities in the circuit and your thinking, and the same action will occur. At one time, most bipolar transistors were pnp devices. but today it is a mixed bag, and the npn probably predominates.

Transistor Gain

There have been several popular ways to denote transistor gain. The two most popular are designated *alpha* and *beta*. The alpha gain can be defined as the ratio of the collector and emitter currents:

$$\alpha = I_c / I_e$$

where I_c is the collector current, and I_b is the base current, expressed in the same units as the collector current.

☐ **Example:**

Find the alpha gain when the collector current is 8 mA and the emitter current is 8.2 mA.

$$\alpha = I_c / I_e$$
$$= (8\,mA)/(8.2\,mA)$$
$$= 0.98$$

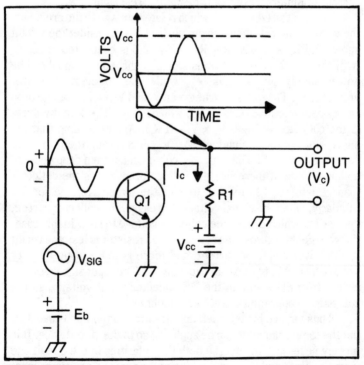

Fig. 2-5. Transistor as a voltage amplifier.

The value of the alpha gain is less than unity (1), with typical values between 0.7 and 0.99.

The parameters just used are static measurements, but a more realistic view is the AC alpha, which is defined as the ratio of a small change in collector current caused by a small change in emitter current:

$$\alpha = \Delta I_c / \Delta I_e$$

☐ **Example:**

A collector current that changes from 23 mA to 23.5 mA is caused by a change in base current of 1.12 mA to 1.64 mA. Find the alpha of this transistor.

$$\alpha = \Delta I_c / \Delta I_e$$
$$= (23.5 - 23.0)/(1.64 - 1.12)$$
$$= (0.5)/(0.52) = 0.96$$

The gain designation preferred for the common-emitter transistor amplifier, the circuit most frequently used, is the *beta gain*, which is defined as the ratio of the collector to the base current:

$$\beta = I_c / I_e$$

or for AC beta:

$$\beta = \Delta I_c / \Delta I_b$$

In some cases, we see the beta designated as H_{fe} for the DC beta and h_{fe} for the AC beta.

☐ **Example:**

Find the beta gain of a transistor in which a base current change of 100 μA causes a collector current to change from 66 to 79 mA.

$$h_{fe} = \Delta I_c / \Delta I_b$$
$$= (79 - 66)/(0.1)$$
$$= (13)/(0.1) = 130$$

Alpha and beta are preferred for different classes of transistor amplifier circuit, but they are mathematically related. Recall that both involve the collector current, so they can be expected to have a relationship to each other. The expressions are:

$$\alpha = \frac{\beta}{1 + \beta}$$
$$\beta = \frac{\alpha}{1 - \alpha}$$

☐ **Example:**

Find the beta gain of a transistor that is known to have an alpha gain of 0.94.

$$\beta = \alpha/(1 - \alpha)$$
$$= \alpha/(1 - \alpha)$$
$$= (0.94)/(1 - 0.94)$$
$$= 0.94)/(0.06) = 15.7$$

☐ **Example:**

Find the alpha gain of a transistor that is known to have a beta gain of 100.

$$\alpha = \beta/(1+\beta)$$
$$= (100)/(1 + 100)$$
$$= (100)/(101) = 0.99$$

FREQUENCY RESPONSE OF TRANSISTORS

The frequency response of transistor devices can be measured using any of three different methods: *alpha cutoff, beta cutoff* and *gain-bandwidth product*. The alpha cutoff frequency, F_{ab}, is defined as the frequency at which the AC alpha gain, h_{fb}, falls off to a value that is 3 dB lower than the alpha gain at some low frequency, such as 100 Hz; i.e., $h_{fb} = (0.707)h_{fbo}$, where h_{fbo} is the current gain at 1000 Hz.

The beta cutoff frequency is defined in a similar manner, and is the frequency at which the beta gain of the transistor falls off to a value that is 3 dB down from its 1000-Hz value; i.e., $h_{fea} = (0.707)$ h_{feo}. In general, the alpha cutoff is higher than the beta cutoff, and that the beta cutoff frequency is a more ideal representation of the performance of a transistor.

The most useful transistor frequency response expression seems to be the *gain-bandwidth* product, f_T. This frequency is defined as the frequency at which the gain falls off to unity (beta gain is used). The gain-bandwidth product if valid for transistors operated in the common emitter configuration. We can express f_T as:

$$f_T = h_{fe} \times f_o$$

where f_T is the gain-bandwidth product, h_{fe} is the AC beta, f_o is the frequency at which the gain is measured. We find the value of f_T listed in transistor specification sheets as the frequency at which h_{fe} drops to unity. In many cases, we can find the gain-bandwidth product of a transistor if we know beta cutoff frequency f_{ae}:

$$f_T = f_{ae} \times h_{feo}$$

The f_t is often approximately equal to or slightly less than the alpha cutoff frequency.

In ordinary amplifier applications of the circuit shown in Fig. 2-5, it is normal to select a value of bias current that will cause a collector current to drop V_c to a point approximately equal to $\frac{1}{2}$-V_{cc}. This will allow an input signal with an amplitude that will exactly double the value of V_c on one peak of the sine wave and drop to exactly zero on the peak of the other half-cycle. A bias voltage must be provided to make the transistor operate at this *quiescent point*.

It is not economical to use separate DC power supplies for both the collector-emitter and base-emitter voltages, especially when the emitter is common to both circuits. A resistor *bias network* can instead derive the base-emitter potential from the collector-emitter power supply.

The simplest form of bias network is shown in Fig. 2-6. In this circuit, bias resistor R_b is connected between the base of the transistor and the +V_{cc} power supply (– V_{ee} if the transistor is pnp). The collector load resistor is connected from the collector to the same power supply. The collector current flows from the collector through the collector load resistor to the V_{cc} power supply. Similarly, base current I_b flows from the base terminal of the transistor, through base resistor R_b and to the V_{cc} power supply. The value of the base current is selected to set the collector-emitter voltage to approximately $\frac{1}{2}$-V_{cc}.

The circuit shown in Fig. 2-6 provides the highest possible voltage gain, but it has terrible thermal stability. Temperature affects the base-emitter voltage of the transistor (Equation 2-1). In this circuit, the thermal affects are maximal. We select some needed, or convenient, level of collector current. Then we set the bias current according to Ohm's law, considering the collector supply voltage, V_{cc}, and the junction voltage required for the transistor (0.6 volts in silicon transistors).

The thermal stability of the circuit can be improved by creating a small amount of negative feedback in the emitter circuit. This is done by connecting a small resistor in series with the emitter (R_e in Fig. 2-7). The resistor works by using the emitter current to create a voltage drop across the emitter resistor. The emitter current is directly related to the collector current, so it too will change with temperature. When the temperature is stable, the voltage drops across the resistor, hence the voltage applied to the emitter terminal of the transistor is constant. But when the temperature changes, this voltage changes also. The bias on the transistor affects the collector current and is defined as the difference between the base and emitter potentials. Because the base voltage is fixed, the change in emitter voltage will effectively change the transistor bias and alter the collector current. Consider

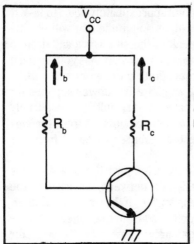

Fig. 2-6. Transistor bias scheme for operating from a single power supply.

the case when the transistor temperature increases. According to Equation 2-1, the collector current will increase at this time, and that will cause an increase in the voltage drop across the emitter resistor, R_e. There is then a higher voltage on the emitter, which

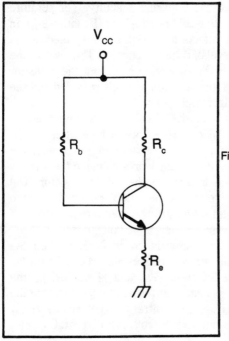

Fig. 2-7. Another bias scheme.

tends to reduce the base-emitter voltage bias of the transistor, cancelling the change in collector current occasioned by the temperature rise. A similar change occurs in the opposite direction when the temperature decreases causing a decrease in the emitter current.

We can also gain a measure of stability using the bias configuration of Fig. 2-8. The emitter resistor is used (although in many cases it is deleted), and the base bias resistor is returned to the collector of the transistor instead of directly to the V_{cc} power supply. At frequencies up to 100 kHz (at higher frequencies, certain reactances enter into the picture) certain generalizations concerning the input impedance, output impedance, current gain and voltage gain can be made. For example, the low frequency, small-signal output impedance is approximately equal to the collector resistor, R_c. The input impedance is approximately the product of emitter resistance R_e and the AC beta of the transistor. For example, assume that a 560-ohm emitter resistor is used in a circuit that has a beta of 90. What is the input impedance? By the rule-of-thumb just given, it would be (560 ohms) (90), or 50,400. The current gain of the stage is approximately the beta rating.

The voltage gain of the stage shown in Fig. 2-8 is the ratio of the collector resistor (R_c) to the emitter resistor (R_e), which is then multiplied by the AC beta:

$$A_v = \frac{R_c}{R_e} \times h_{fe}$$

where A_v is the voltage gain, R_c is the collector resistance in ohms, R_e is the emitter resistance in ohms, and h_{fe} is the AC beta.

In practical circuits we must make certain assumptions about the relative values of the emitter and collector resistances. In most small-signal circuits—not in power amplifiers—we will have an emitter resistor in the 100-ohm to 2000-ohm range, with most being in the under-1000-ohm range. Also, we find that some trade-off must be made with respect to the relative values of these resistors. If the emitter resistor is a small percentage of the collector resistor (0.05 or less), the thermal stability of the circuit suffers, but the voltage gain is improved. Similarly, if the emitter resistor is made a large percentage of the collector resistor, the thermal stability is improved, but the voltage gain is lowered. The ratio of resistances must be adjusted for the most desirable trade-off between stability and gain. In general, the emitter

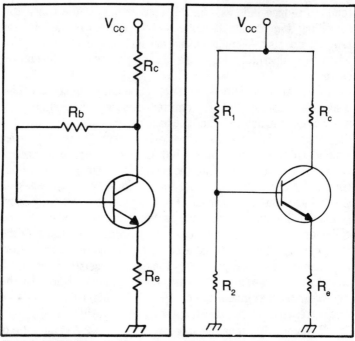

Fig. 2-8. Another bias scheme.   Fig. 2-9. Another bias scheme.

resistor will have a value that is between 10 percent and 20 percent of the collector resistance.

The circuit of Fig. 2-8 achieves its stability from the emitter resistor and the configuration of the base resistor connection. Consider the action of the collector voltage on the base current, as temperature changes force changes in the collector current. For the moment, assume that the emitter resistance is zero.

Perhaps the best bias circuit is the fixed-bias arrangement of Fig. 2-9. In this circuit, a resistor voltage divider fixes the base voltage at some specific point. Again, at low frequencies, the output impedance is approximately equal to the collector resistance, but there are some differences in the other parameters. The input impedance is approximately equal to the base resistor that is connected to ground, which is the resistance of R2 in Fig. 2-9. The current gain of the stage if given by the ratio of the grounded resistor in the base and the emitter resistance, or R_2/R_e. The voltage gain is approximately equal to the ratio of the collector and emitter resistors. As a rule-of-thumb, the value of R2 should be approximately 10 times the value of the emitter resistor.

32

Chapter 3
Amplifier Basics

Certain concepts are common to all amplifiers, and still others are common to large classes of amplifiers using all forms of active devices as the amplifying elements. The classification of amplifiers as to the action element is well known. It is easy, for example, to see why some amplifiers are vacuum-tube amplifiers and others are transistor amplifiers. Transistor amplifiers are even broken down into bipolar (npn and pnp) and field-effect (JFET and MOSFET) types. But we also have several other methods of classifying amplifier circuits, which are not so dependent upon the technology of the active device. Several of these classification schemes are:

- ☐ Classification by common element
- ☐ Classification by conduction angle
- ☐ Classification by transfer function
- ☐ Classification by feedback versus nonfeedback
- ☐ Classification as to frequency DC, low-frequency AC, wide-band or rf.

In this chapter we will consider some of the basics of amplifier circuits. Keep in mind that any of these could be constructed using transistors of either basic type or tubes. They can also use linear integrated circuits.

CLASSIFICATION BY COMMON ELEMENT

This method of classifying amplifier circuit revolves around noting which element (cathode/grid/plate, base/emitter/collector or source/drain/gate) is common to both input and output circuits. Although technically incorrect, this is sometimes referred to as the *grounded* element, such as grounded-emitter amplifiers. We tend to use *common* and *grounded* interchangeably, so bear with us if you are a purist.

Consider this topic from the point of view of the transistor. We can extend the concept to tubes and FETs by keeping in mind the analogy of elements shown in Table 3-1. So when the text states grounded base or common base, you can extend the same concept to vacuum tubes (grounded grid) and FETs (grounded gate) by substituting the analogous element.

Figure 3-1 shows the different entries into this class. The circuit shown in Fig. 3-1A is the common emitter circuit. It gets its name from the fact that the emitter terminal of the transistor is common to both input and output circuits. The input signal is applied to the transistor between the base and *emitter* terminals, while the output signal is taken from across the collector and *emitter* terminals. Hence, the emitter is common to both circuits.

The common emitter circuit offers high-current amplification—the beta rating of the transistor. But the common emitter also offers a substantial amount of voltage gain, as well. Recall from Chapter 1 that the transistor is also a voltage amplifier, especially when a series resistor is placed between the collector terminal and the collector power supply. The values of gain for current and voltage are vastly different. The current gain is h_{fe}, but the voltage gain depends upon other factors. In the next chapter, you will see that the voltage gain depends upon the R_L/R_E ratio in some cases and the product of the ratio and the beta in other cases.

The input impedance of the common emitter amplifier is medium ranged, or in the 1K ohm range. The output impedance, though, is typically high, up to 50K ohms. Typical values will be determined by the specific type of circuit, but there are some approximations that can be made. For most common emitter circuits, Z_{in} is equal to the product of the emitter resistor (R_E) and

Table 3-1. Analogous Parts of Three Amplifier Devices.

Bipolar Transistor	Vacuum Tube	Field-Effect Transistor
Base	Grid	Gate
Emitter	Cathode	Source
Collector	Anode (plate)	Drain

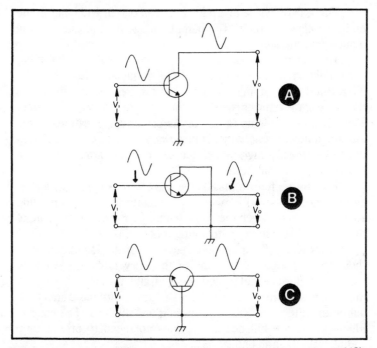

Fig. 3-1. (A) common-emitter amplifier, (B) common-collector amplifier and (C) common-base amplifier.

the h_{fe} of the transistor. The output impedance is essentially the value of the collector load resistor and will range from 5K to 50K ohms.

The output signal in the common emitter circuit is 180 degrees out of phase with the input signal. This means that the CE amplifier is an *inverter* stage. The output signal will be negative-going for a positive-going input signal, and vice versa. The common-emitter transistor amplifier—and its tube/FET analogs—is probably the most often used configuration. It is certainly the most often cited in textbooks.

Common Collector

This configuration is shown in Fig. 3-1B. In this circuit, the collector terminal of the transistor is common to both the input circuit and the output circuit. This circuit is also called the *emitter follower circuit.* The common collector circuit offers little or no voltage gain. Most of the time the voltage gain is actually less than unity. But the current gain is considerably better, such as $h_{fe} + 1$.

There is no phase inversion between output and input in the emitter follower circuit. The output voltage is in phase with the input signal voltage.

The input impedance of this circuit tends to be high, sometimes greater than 100k ohms at frequencies less than 100 KHz. But the output impedance is very low, because it is limited to the value of the emitter resistor, which can be as low as 100 ohms. This situation leads us to one of the primary applications of the emitter follower: *impedance transformation*. The circuit is often used to connect a high-impedance source to an amplifier with a low-impedance amplifier.

The emitter follower is also used as a *buffer* amplifier, which is an intermediate stage between two circuits in order to provide some isolation between the two circuits. One primary example of this application is in the output circuit of oscillator circuits. Many oscillators will pull, or change frequency, if the load impedance they drive changes. Yet some of the very circuits used with oscillators (a superhet receiver) naturally provide a changing impedance situation. The oscillator will prove a lot more stable if a buffer amplifier is provided between it and the load. The emitter follower fills the bill because it does not need to provide any additional voltage amplification.

Common Base Circuits

Common base amplifiers use the base terminal of the transistor in both the input and output sections of the amplifier. The input signal is applied between the emitter and the base (see Fig. 3-1C), and the output is taken between the collector and the base.

The voltage gain of the common base circuit is high, on the order of 100 or more; however, the current gain is low, usually less than unity. The input impedance is also low, usually less than 1000 ohms, because it is limited by the emitter resistor. On the other hand, the output impedance is quite high. Again there is no phase inversion between input and output circuits.

The principal use of the common base circuit is in high-frequency (HF) and VHF/UHF amplifiers in receivers. The circuit requires no neutralization at these frequencies, so it is superior to the common emitter circuit. Neutralization prevents oscillator action due to interelectrode capacitances providing a feedback signal that is in phase with the input signal. Table 3-2 summarizes the properties of the common emitter, common base and common collector transistor amplifier circuits.

CLASSIFICATION BY CONDUCTION ANGLE

When one speaks of amplifier classes, it almost always means classification by *conduction angle*, the portion of the input cycle over which output current flows. We recognize three traditional classes, labeled A, B and C, and a combination class, AB. A new class is sometimes called class-D.

□ Class-A: In this class, collector (plate/drain) current flows over the entire 360 degrees of the input cycle. This class is the least efficient of the three, but it is capable of linear (low distortion) operation using just one transistor.

□ Class-B: In this class, collector current flows over 180 degrees of the input cycle. This class is more efficient that class-A but requires two or more transistors in order to make the amplifier linear.

□ Class-C: This class is the most efficient of the classes and has collector current flowing over less than 180 degrees of the input cycle. A typical conduction angle for a class-C amplifier is 90 to 150 degrees, with 120 degrees being most common.

The efficiency of an amplifier is dependent in part upon the class of operation. Figure 3-2 shows a chart of efficiency versus conduction angle. In this case, efficiency is defined as the ratio of the collector DC input power to the signal output power, expressed as a percentage

$$n = \frac{P_o}{V_c I_c} \times 100\% \qquad \textbf{Equation 3-1}$$

where n is the efficiency expressed as a percentage, V_c is the collector voltage (rms), I_c is the collector current (rms), and P_o is the output power in watts.

□ **Example:**

Calculate the efficiency if an amplifier delivers 250 mW into a load when the collector current is 80 mA and the collector potential is 12V.

$$n = (P_o)(100\%)/(I_c V_c)$$
$$= (0.25)(100\%)/(12 \times 0.08)$$
$$= (0.25)(100\%)/(0.96) = 26 \text{ percent}$$

The class-A amplifier does not operate at a great efficiency. In fact, most operate at 25 percent efficiency. These amplifiers, then, would consume large amounts of DC power in order to generate any appreciable amount of signal output power. As a result, class-A

Table 3-2. Characteristics for the Three Basic Amplifier Configurations.

	Common Emitter	Common Collector	Common Base
Voltage Gain	High	Low (≈ 1 or less)	High (>100)
Current Gain	High	High ($\beta + 1$)	Low (<1)
Z_{in}	Medium (≈ 1 K)	High (>100 K)	Low
Z_{out}	Medium to High (50 K)	Low ($<100\ \Omega$)	High (>500 K)
Phase Inversion?	Yes	No	No

amplifiers are used almost exclusively as voltage amplifiers, which deliver little actual power but do build up the power level considerably. One exception to the rule is in situations where

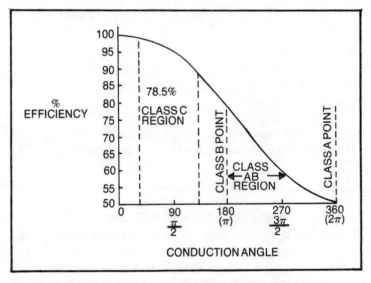

Fig. 3-2. Efficiency-versus-operating angle for the four amplifier classes.

power is plentiful and component count is costly. In automobile and home broadcast receivers, for example, the designer worries little about the power supply but is terribly concerned with the overall parts count in the radio. As a result, many have used class-A power amplifiers despite the low efficiency. Almost all American-made auto radios since 1962 have used a single-ended, class-A, audio power amplifier in the output stage.

The lost power—the difference between the DC power consumed and the signal output power delivered to an external load—is given off as heat in the transistor. The collector temperature of a transistor is considerably higher when operated in class-A to deliver any given amount of output power. Also, at zero signal, when there is no input signal, the collector dissipation is maximum. In the class-B and class-C designs, on the other hand, the collector dissipation drops to zero when the input signal is zero.

The class-B amplifier develops collector power only when the input signal is not zero. This is because the collector current flows over only one-half, or 180 degrees, of the input cycle. But it is impossible to make a linear amplifier using just one transistor in class-B. It takes *two* transistors, driven out of phase with each other so that they will operate on alternate halves of the input signal. This theme is developed further in Chapter 7 on power amplifiers.

The class-C amplifier cannot be made linear, regardless of the number of transistors used. As a result, the class-C amplifier cannot be used in places where it is important to preserve the input waveform. There are no class-C hi-fi amplifiers, for example. Similarly, there are no radio frequency power amplifiers in class-C if single sideband or amplitude-modulated signals are to be boosted. The class-C amplifier is, however, frequently used in radio power amplifiers if the transmitter is for CW (continuous wave), frequency modulation or high-level amplitude modulation, in which the modulating signal is applied to the final amplifier plate circuit.

Rf amplifiers operated in class-C might be expected to produce a high harmonic output—not nice in radio transmitters! But the high harmonic content of the raw pulse signal is eliminated by the action of the rf tank circuits. Pulsed tank circuits will oscillate at their natural resonant frequency. If the tank circuit is adjusted to have the same frequency as the plate/collector pulses, the output will be "resinerized" by the flywheel effect. Class-C efficiency is very high, typically on the order of 70 to 80 percent.

The class-AB amplifier circuits are an attempt, not without success, to obtain some of the benefits of both classes. When two class-AB transistors are used in push-pull, the circuit will prove more linear than the class-B and more efficient than the class-A.

CLASSIFICATION BY TRANSFER FUNCTION

The transfer function of any electronic circuit can be expressed as the ratio of the output to the input. For a voltage amplifier, for example, the transfer function is V_o/V_{in}, which results in the *voltage gain* of the circuit (symbolized by A_v).

There are four general subclasses of transfer function, and these are called: voltage, current, transconductance and trans-resistance. Most readers are already familiar with the first two, voltage and current. A voltage amplifier has a transfer function that expresses the ratio of the output voltage to the input signal voltage:

$$A_v = V_o/V_{in}$$

The units of voltage gain A_v are dimensionless because the *volts* terms in both numerator and denominator cancel each other.

The current amplifier is given a transfer function that expresses the current gain, A_i, as a function of the output current and input signal current:

$$A_i = I_o / I_{in}$$

These two forms of amplifier are very well known and are understood by most readers. But what about transconductance and transresistance amplifiers? These are not quite as well known, except among engineering and technology students. Some hobbyists and amateur radio operators, for example, are totally unaware of these amplifier types, even though they see them repeatedly.

The transresistance amplifier has a transfer function that relates a output voltage to an input current. The units *resistance* apply to the transfer function because in Ohm's law, V/I is a resistance. The transfer function, then, is:

$$R_m = V_o / I_{in}$$

The resistance of the transfer equation is measured in the ordinary units of the ohm.

The transconductance amplifier is a little more familiar to most, because most ordinary vacuum tubes and field-effect transistors are described in terms of transconductance. The unit of transconductance is merely the reciprocal of resistance, so we expect a transfer function of the form I/E. The unit used to describe transconductance is the *mho* (which is ohm spelled backward). Any transconductance amplifier will describe an output current caused by an input voltage:

$$g_m = I_o / V_{in}$$

☐ **Example**

Find the transconductance in micromhos of a field-effect transistor amplifier in which a 500 mV change in signal voltage will cause a 1-mA change in output current.

$$
\begin{aligned}
g_m &= I_o / V_{in} \\
&= (0.001A)/(0.500V) \\
&= 0.002 \text{ mhos} \times (\ 10^6\ \mu\text{mhos/mho}) \\
&= 2000\ \mu\text{mhos}
\end{aligned}
$$

CLASSIFICATION BY FEEDBACK VERSUS NONFEEDBACK

Thus far we have not mentioned the concept of feedback in amplifiers very seriously. But feedback can make a mediocre amplifier act like it is a much higher quality circuit. It is possible to cancel some distortion in the amplifier and make it a lot more stable by using a little *negative feedback*. In this section, we will deal briefly with the concept of negative feedback.

The basic block diagram for a feedback amplifier is shown in Fig. 3-3. The main amplifier will be a transistor, IC or vacuum-tube stage and will have a gain A. The output signal voltage, V_o, is sampled, and this sample is passed back to the amplifier input via a feedback network. The gain of the feedback network is given by the Greek letter *beta*, β. In most cases, the gain of the feedback network will be negative, as the network is made of all passive components (resistors, capacitors, inductors, etc.) and no amplification is provided. Of course, it is possible to make a feedback amplifier in which there is a main amplifier stage and a feedback amplifier stage.

The feedback signal is summed with the source signal V_s to form an input signal, $V_i = V_f - V_s$. The output signal is the product of this input signal and the voltage gain of the stage. In this section we are dealing with a voltage amplifier system but could just as easily call the circuit any of the other types of amplifiers by substituting the correct parameters.

The gain of the amplifier *with feedback* is given by the expression:

$$A_v = \frac{A}{1 + A\beta}$$

where A_v is the gain of the amplifier with feedback, A is the gain of the amplifier without feedback, and β is the transfer gain of the feedback network.

☐ **Example:**

Calculate the voltage gain of an amplifier when forward voltage gain A is 1000 and the feedback factor is 0.01.

$$A_v = A/(1 + A\beta)$$
$$= (1000)/(1 + (1000)(0.01))$$
$$= (1000)/(1 + (10)$$
$$= (1000)/(11) = 90.9 = 91$$

The use of feedback will improve the distortion situation of the amplifier. The *total harmonic distortion* (THD) will be reduced in the feedback amplifier, as in high-fidelity amplifiers. Very few audio amplifiers intended for high-fidelity applications do not have rather extensive feedback. The degree of feedback is often expressed in decibels:

$$N_{dB} = 20 \log_{10}(A_v/A) = 20 \log_{10}(1/(1 + A\beta))$$

Figure 3-4 shows an example of a feedback network involving only resistors. We will use this as an example of a simple feedback network to show how the factor, β, is derived for this case. The feedback network is a simple voltage divider using two resistors, R1 and R2. The value of beta is given by the transfer function of the feedback network. Recall that any transfer function is merely the output voltage divided by the input voltage. In this case, the *output voltage* is feedback voltage V_f, and the *input voltage* is the amplifier output voltage, V_o. If this seems confusing, consider that the *is a* feedback network! From our knowledge of elementary voltage divider theory, we know that

$$V_f = \frac{V_o R1}{R1 + R2}$$

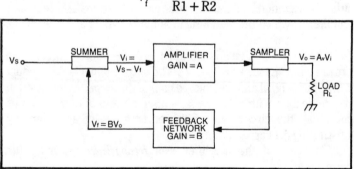

Fig. 3-3. Block diagram of a feedback amplifier.

The transfer function is obtained by dividing each side of the equation by V_o:

$$\frac{V_f}{V_o} = \frac{R1}{R1 + R2}$$

Because beta is defined as V_f/V_o, we can state that

$$\beta = \frac{R1}{R1 + R2}$$

From this equation and the previous example, you can see that it is possible to set the gain of the amplifier by merely manipulating the resistors in the feedback network! This fact will not be lost on many of us once we get into the chapter on operational amplifiers. This ability to easily set the voltage gain of the overall amplifier is one of the principal advantages of the operational amplifier (op amp).

CLASSIFICATION BY FREQUENCY RESPONSE

Amplifiers are also classified as to the approximate frequency response. This method is almost anecdotal, and there is considerable overlap of the various classes. One class is a *DC amplifier*. This type of amplifier will pass all AC frequencies and DC levels. The fact that it is a DC amplifier does not mean that it won't pass AC, but that *will* pass DC signals as well as AC signals.

An *audio amplifier* is one that will generally pass AC signals in the audio frequency range of roughly 30Hz to 20,000 Hz. A communications audio amplifier may well have a much more limited bandwith, such as 300 Hz to 3000 Hz. Hi-fi amplifiers, on the other hand, often have upper-end frequency responses into the 100 kHz range, yet are still called audio amplifiers.

A *wideband amplifier* is exactly what its name implies: It presents a wide bandwidth. You might find a wideband amplifier presenting a bandwidth of dozens or hundreds of kilohertz. An amplifier that has a response up to 100 or 1000 kHz is surely a wideband amplifier. We would also call an amplifier with a 100mHz bandwith "wideband," but it is the practice of some to call amplifiers with bandwidths of over a couple of megahertz *video amplifiers*.

Rf amplifiers are usually tuned to some specific frequency in the rf range. These amplifiers are used in radio communications (receivers and transmitters) to select only the frequency of interest. These amplifiers might be wideband but will be centered about the frequency of interest. Most, however, are essentially narrowband devices.

AMPLIFIER COUPLING METHODS

It is usually necessary to provide some coupling between stages in cascade. When you want to pass a signal from one stage to the next, you must make sure that the proper conditions are met. For example, in some power amplifier applications the coupling circuit must also match impedances between the stages (maximum power transfer occurs when the impedances of the source and load are equal). In other cases, we are not interested so much in matching impedances but are critically interested in keeping DC voltages from the input stage from interfering with the operation of the stage. In transistor amplifiers, for example, the collector voltage may well be 10V to 40V, while the base-emitter voltage of the next stage will be in the 0V to 5V range. Clearly, when these stages are directly connected, it is likely that trouble results (talk about understatement!). But there is, however, a certain type of coupling circuit in which the transistor elements are connected together: *direct coupling*.

Direct Coupling

When direct coupling is used, the collector of the input stage is connected to the base of the next stage in cascade. It is necessary to design these stages such that the base voltage of Q2 (Fig. 3-5) is the same as the collector voltage of Q1.

One advantage of the direct-coupled amplifier is that it will pass signals at all frequencies down to and including DC. These amplifiers are then sometimes called DC amplifiers, not after the name "direct coupled" but because they will pass DC signals.

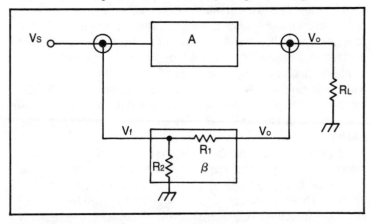

Fig. 3-4. Feedback amplifier with resistor feedback network.

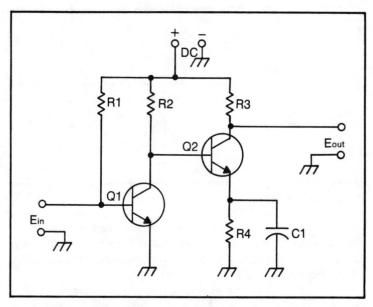

Fig. 3-5. Direct-coupled amplifier.

There could theoretically be almost any number of stages connected in this manner, but in most cases the amplifier will have only two to 10 stages. Otherwise, you run into the supply voltage limit and there is no longer any swing available for the signal.

Transformer Coupling

Figure 3-6 shows an example of transformer coupling. This particular circuit shows two transformers. T1 is an interstage

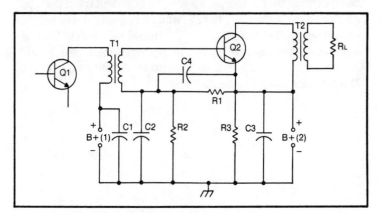

Fig. 3-6. Transformer coupling amplifier.

45

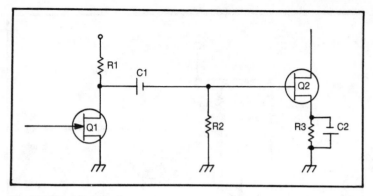

Fig. 3-7. Capacitor coupling.

transformer and T2 is an output transformer. The principal reason for this type of coupling is impedance matching, so it is used mostly in power amplifier applications. On any transformer, the ratio of the impedances is set by the turns ratio of the windings:

$$(Z_p/Z_s)^{1/2} = (N_p / N_s)$$

where Z_p is the impedance seen by the primary (the reflected impedance across the transformer), Z_s is the impedance of the load connected to the secondary winding, N_p is the number of turns in the primary, and N_s is the number of turns in the secondary.

Capacitor Coupling

Figure 3-7 shows an example of resistor/capacitor coupling. The idea here is to pass the signal from the input stage (Q1) to the output stage (Q2) without letting the DC voltage used to operate the input stage bias the output stage. The drain voltage of the input JFET can be as high as 25V to 40V, while it is almost a sure bet that the voltage applied to the gate of Q2 should be not only low, but negative. Both the level and the polarity, then, are wrong. Capacitor C1 is used to keep the DC that is present on the drain of Q1 from being applied to the gate of Q2.

Chapter 4
Designing Transistor Amplifiers

Before you can apply transistors properly you'll need to know some of the rules of thumb regarding designing practical transistor amplifiers. The actual process of design is given in some greater complexity in other books, because they are dealing with generalized cases that can be applied to a wide range of situations. But we are discussing here the standard class-A amplifier, in which the output is a voltage that is a linear reproduction (except for amplitude) of the input signal.

TRANSISTOR RATINGS

The manufacturer's specifications sheet or data book will list certain DC specifications for each different transistor type number. We will see specs for maximum collector voltage, maximum collector current and maximum collector dissipation. When you design a transistor amplifier, you must be cognizant of these matters. The *maximum collector voltage* is the maximum potential that can be applied between the collector and emitter terminals of the transistor without causing permanent damage to the device. Similarly, the *maximum collector current* is the maximum value of I_c that can be sustained without causing damage.

There is little trouble regarding these maximum ratings until you try to operate the transistor at some values close to both maximums. It is quite frequently the case that the product of the maximum collector voltage and the maximum collector current will exceed the maximum allowable collector dissipation! While this would still be well within the maximum current and power ratings, it is likely that this would exceed the maximum power dissipation. Be careful to check the maximum ratings under all circumstances!

LOAD LINES

We want to vary the base current of the transistor in a manner that will cause the collector current to vary also, and in a linear manner. The transistor manufacturer will make available data called a family of curves, which relates the I_c-versus-V_c levels for assorted values of base current (Fig. 4-1). Each curve in the family represents the collector current caused by any given collector voltage for specified levels of base current. In this case, base bias levels of 100 μA, 200 μA, 300 μA, 400 μA and 500 μA have been selected.

The load line is drawn onto the family of curves by noting the points at which the collector current is zero and maximum. The load line is drawn between these two end-points. The collector current is zero (neglecting the leakage current that invariably exists) when the transistor is cut off. In most amplifiers, a collector load resistance is in series with the DC collector power supply. Under conditions when the collector current is zero, the voltage drop across this resistor is also zero, so the collector-emitter voltage of the transistor is equal to the supply potential, V_{cc}. The current is limited by the collector load resistor, so it will assume a value equal to V_{cc}/R_c under conditions where the transistor is saturated. Figure 4-1 shows a 1000-ohm load line, which is the load line when the collector resistor is 1K ohm. When the collector current is zero, the output voltage V_c rises to the full 28V delivered by the power supply. And when the transistor is saturated, the output voltage is zero, and the collector current is maximum. All of the voltage from the power supply is dropped across the collector resistor, so the collector current is 28 mA (28V/1000 ohms).

The *quiescent point* is the point defined when the input signal is zero. In most transistor, class-A amplifiers, the quiescent point will be the condition in which the collector voltage, V_c, is equal to $\frac{1}{2}V_{cc}$, or in our present case 14 VDC. In this example the quiescent collector current will be 14 mA. When an AC signal is applied to the base of the transistor, it will modulate the base current up and down about the quiescent point, causing the collector current to vary. Voltage amplification is obtained because the varying collector current will cause the percentage of V_{cc} that is dropped

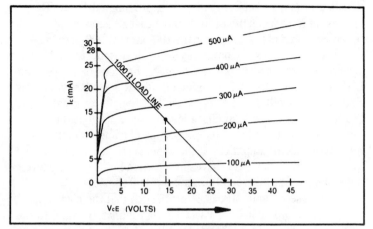

Fig. 4-1. Transistor operating curves with load line.

across R_c to vary accordingly. One goal designers is to choose values for these parameters that will achieve the correct results, and then bias the transistor properly.

Perhaps the best way to demonstrate some of the principles is by example. Figure 4-2 shows the first of our simple transistor amplifiers. It is the simple bias network shown previously. In this case, there is an emitter resistor. This resistor improves stability under temperature variations and increases the input impedance of the amplifier. The value of the collector resistor is used to set the quiescent point to which we will bias the transistor. There are several different criteria for the value of the collector resistor. In some cases, for example, you might want to match some impedance and will therefore select the collector resistor accordingly (the output impedance of this stage is approximately R_c). In other cases you will merely select a value that will place the transistor in the middle of its operating range. In most cases, the designer will pick a value of collector current that is deemed desirable, and the power supply voltage will be selected either by constraints of the design or the equipment in which the amplifier is used. As often as not, the value of the collector voltage is determined solely by the available power supply. In some instances, however, you might wish to achieve some specific output voltage swing and will select the power supply voltage for this value.

The emitter resistor value is always a trade-off or compromise between thermal stability and gain. When emitter resistor R_e is made larger, the thermal stability is improved, but this improvement in stability is only at the expense of lost voltage gain.

Recall the gain of this amplifier is approximated by R_c/R_e, so increasing R_e reduces the ratio. In most cases, the value of R_e will several hundred ohms, at least in the 100-ohm to 2000-ohm range.

The value of R_e is obviously of critical importance. It has become almost standard practice to make the value of this resistor 1/10 to 1/5 of the value selected for R_c. Under very few circumstances will the value of the emitter resistor be greater than $1/5$-R_c. One of these is in phase inverter circuits, but they are a special case that combines the properties of common-emitter and common-collector amplifiers at the expense of circuit gain. In the transistor whose family of curves in shown in Fig. 4-1, the emitter resistor should be between 100 ohms and 200 ohms.

The base resistor, R_b, is selected such that the base current under zero signal conditions will be approximately equal to the quiescent current. In the example of Fig. 4-1, the quiescent base current is 250 μA, and the quiescent collector current is 14 mA. The emitter voltage will be equal to $(0.014A)(R_e)$, so if an emitter resistor of 100 ohms is selected, the emitter voltage is 1.4V. There is a 0.6V base-emitter voltage for silicon transistors, so the total base voltage will be $V_e + V_{be}$, or 1.4V + 0.6V x 2.0 VDC. The DC power supply is 28 VDC, so the base resistor must drop a total of 28V − 2.0V, or 26V, when delivering the quiescent base current of 250 μA. The value of R_b, then, by Ohm's law is:

$$R_b = (V_{cc} - V_b)/I_{bq}$$
$$= (28V - 2V)/(0.00025A)$$
$$= (26V)/(0.00025A) = 104,000 \text{ ohms}$$

In most cases, you would not try to find the value resistors called for in the calculations but would attempt first to make the circuit operate with the nearest standard value, in this case 100K ohms.

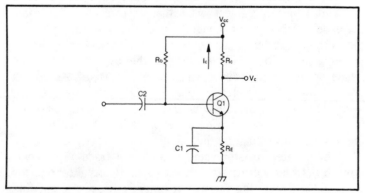

Fig. 4-2. Simple transistor amplifier.

Another version of the circuit shown in Fig. 4-2 connects the base resistor directly to the collector terminal, rather than to the power-supply potential. This strategy gives us better thermal stability because of some degenerative feedback present. The method for selecting the value of the base resistor is exactly the same as in the previous case, except that the voltage used will be the collector voltage, V_c, instead of V_{cc}. This will be approximately one-half of the power-supply voltage in most applications.

The emitter bypass capacitor is used to keep the emitter at a low impedance to ground for AC frequencies. The emitter bypass capacitor should have a reactance that is one-tenth of the resistance of R_e at the lowest frequency of operation. In the example, we used a emitter resistor of 100 ohms. Suppose that this amplifier was a speech amplifier in a two-way radio. The audio frequency response has a low-end limit of 300 Hz in most such equipment, so a capacitor that will present a reactance of 100/10, or 10 ohms, at 300 Hz must be selected. From the standard equation for capacitive reactance, this value is 53 μF, or more. Fortunately, the voltage at this point is low—approximately 4 volts under maximum collector current conditions—so almost any small-size electrolytic capacitor will work fine.

The circuit of Fig. 4-2 suffers from several defects, so it is not quite as popular as the circuit of Fig. 4-3. In this more complex circuit, the base voltage is set by the resistor voltage divider in the base circuit. The collector and emitter circuits are the same as in the other case. Once again, the output impedance is the collector load resistor. The input impedance, however, is limited by the value of R2. The design procedure is simplified below:

☐ Select an appropriate transistor type. It is absolutely essential that you know the type of transistor used, have a data sheet on that transistor and are reasonably confident that the selected transistor will do the job.

☐ Select appropriate values of I_c, I_b and V_c. The value of I_c may almost be arbitrary, and the collector voltage will be set by the power-supply constraints.

☐ Select a value for the voltage on the emitter. In most class-A amplifiers, the emitter voltage will be roughly one-tenth of V_c. Check the selections to make sure that the emitter voltage is compatible with the normal range of emitter resistances when the selected value of collector current flows.

☐ Select a value for R_c. This resistor will be $(V_{cc} - V_e - V_c)/I_c$.

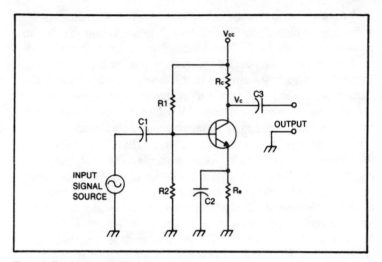

Fig. 4-3. Transistor amplifier.

☐ Select a value for R_b (a combination of R1 and R2). A rule of thumb is used here, namely select a value that is 10 to 20 times R_e. Most will use $15R_e$ for R_b.

☐ Calculate V_{bb} from V_b. A 0.6 volt junction potential is across b–e, and a voltage is on the emitter called, V_e. The value of V_b is then $V_e + 0.6$.

$$V_{bb} = V_b + I_b R_b$$

☐ Select R1 and R2 from the following expressions:

$$R1 = R_b(V_{cc}/V_{bb})$$

$$R2 = (R_b R1)/(R1 - R_b)$$

The parameters R_b and V_b are not real but are used analytically to arrive at the correct values for R1 and R2 when the resistor network is connected directly to the V_{cc} power supply. This method is not foolproof, but it is *approximate*. It will yield positive results for most applications.

Chapter 5
Field-Effect Transistors

The field-effect transistor was the matter of speculation in the late 1930s. A paper published in a scientific journal once gave details of the construction of the function field-effect transistor, and that transistor would have worked if the metallurgy had been developed by that time to provide the semiconductor materials needed. Following World War II, it was the bipolar transistor that received research attention. If the JFET had actually been built in 1939, then the semiconductor age would have had a 10-year headstart on itself, and most transistors today might be field-effect transistors instead of bipolar transistors.

FETs have a very high input impedance and are transconductance amplifiers. They tend to have performance properties that are far more reminiscent of the pentode vacuum tube than actual transistors.

There are two basic types of field-effect transistor: junction field-effect transistors (JFET) and metal-oxide semiconductor FETs (MOSFETs). A basic JFET is shown in Figs. 5-1 and 5-2. The JFET consists of a semiconductor channel and a gate structure. Two basic types of JFET are used: p-channel and n-channel. These designations are based upon the type of material used to form the channel. The gate structure is always made of the opposite type of material. In normal operation, a voltage is applied across the channel in the polarity shown. Some JFETs, incidentally, are capable of operation when the drain and source ends of the

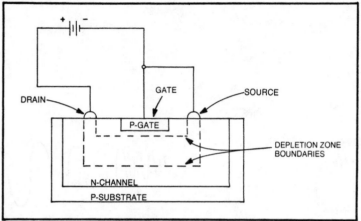

Fig. 5-1. Unbiased junction field-effect transistor structure.

channel are reversed. The channel appears to have a resistance so current flows. When a potential is applied to the gate, however, it will create a *depletion zone* is the channel in which no electrical charge carriers (electrons or holes) can exist. This region has a high electrical resistance. At zero volts gate potential, this depletion zone is at a minimum, so the channel resistance is the lowest. But as the potential is increased, the depletion zone widens, thereby narrowing the portion of the channel in which

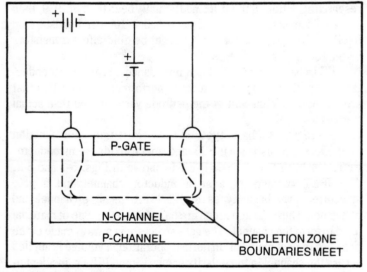

Fig. 5-2. JFET structure.

charge carriers will flow. At some specific potential, called the *pinchoff voltage*, the depletion zone completely chokes off the channel, and no current flows from drain to source.

JFETs

The circuit in Fig. 5-1 is for an n-channel JFET. The gate structure, made of a different form of semiconductor material from the channel, is diffused into the channel in a manner resembling the emitter of a bipolar transistor. But this junction is kept perpetually reverse biased; in fact, if the junction were to become forward biased, destruction of the device would probably result. The channel is formed such that it is placed between the p-type gate and a p-type substrate on which the transistor is built.

Electrons in the n-channel are encouraged to flow by the external battery. If the depletion zone is wide, then current will flow in the external circuit. By making the gate-substrate negative with respect to the channels, electrons are repelled (like charges repel) in the channel, thereby widening the depletion zone. When the depletion zones from gate and substrate meet in the channel, pinchoff occurs. This is shown in Fig. 5-2.

Consider the channel resistance for a moment. When the gate voltage is zero, the depletion zones are narrow, so the channel resistance is low. When a high negative potential is applied to the gate and substrate, however, the depletion zones widen, causing the channel resistance to increase. When the gate potential reaches a certain level, called the *pinchoff voltage*, the channel resistance becomes extremely high. The JFET, then, can be used as a switch because it possesses a high resistance under one condition and a low resistance under the other.

The example in Fig. 5-1 was an n-channel JFET. The p-channel device is exactly the same, except for polarity. The channel structure will be made of p-type semiconductor material, while the gate and substrate will be made from n-type material. The circuits for the p-channel device are merely the n-channel circuits with the external voltage polarities reversed. Symbols for the two devices are shown in Fig. 5-3.

MOSFET devices are a little different. These devices are capable of much high input impedances than JFETs because there is no physical connection between the gate and the channel. In the JFET, input (gate) impedance is limited by the reverse leakage current across the reverse-biased gate-channel junction. There are two basic types of MOSFET device: *depletion* and *enhancement*.

These different devices operate in somewhat different modes and will therefore be discussed separately.

Both types of MOSFET device are available in p-channel and n-channel versions. The possible MOSFETs are:

p-channel Depletion   p-channel Enhancement
n-channel Depletion   n-channel Enhancement

MOSFET devices are also sometimes called *insulated gate field-effect transistors* (IGFET), a name that more nearly describes their construction. The depletion type MOSFET is shown schematically in Fig. 5-4. As in the JFET, the channel will be of one of the two types of semiconductor material (in this case n-type), and the substrate will be made from the opposite material. It is always desirable to keep the substrate-channel pn junction reverse (or at least zero) biased to prevent an excessive current flow. The gate is merely a metallic contact insulated from the channel material by a thin (very thin) layer of metal-oxide insulating material. This is the origin of the name, "insulated gate." The depletion-type MOSFET is a normally-on device, meaning that the channel resistance is low when the gate voltage is zero.

When an electrical potential is applied to the gate, however, an electric field is created in the channel. If the voltage is positive

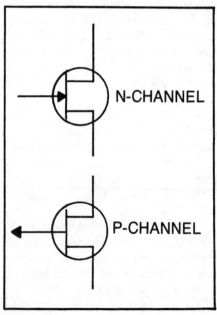

Fig. 5-3. Symbols for p-channel and n-channel JFETs.

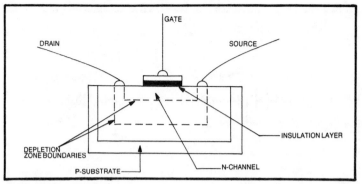

Fig. 5-4. MOSFET structure.

with respect to the channel, then electrons will tend to draw closer to the gate structure. But if the voltage is negative, the field will repel the electrons in the channel, creating a depletion zone. The higher the gate voltage, the wider the depletion zone. When the depletion zone is finally wide enough to pinch off the channel, current flow ceases. We can, therefore, control the channel resistance with a gate voltage.

Field-effect transistors are said to be transconductance amplifiers. The definition of such an amplifier is that a changing voltage at the input controls a change in current at the output. Because the channel resistance is controlled by the gate voltage, the output current will also be under control of the gate, assuming that the drain-source voltage is constant.

The symbol for the depletion MOSFET is shown in Fig. 5-5. As in the case of the JFET devices, the arrow pointing inward means n-channel, and arrow pointing outward means p-channel. It is interesting to note, incidentally, that the manufacture of p-channel depletion transistors is somewhat more difficult than the manufacture of n-channel devices. As a result, the n-channel depletion MOSFET predominates.

ENHANCEMENT MOSFETs

The enhancement MOSFET is a normally-off device (exactly the opposite of the depletion type) in which it is necessary to apply a gate potential before channel construction takes place. Again, both p-channel and n-channel devices are possible. The p-channel enhancement MOSFET requires a negative gate potential to begin conduction, while the n-channel requires a positive gate potential.

Figure 5-6 shows the construction of the enhancement MOSFET. In this case, an n-channel device is shown. In all

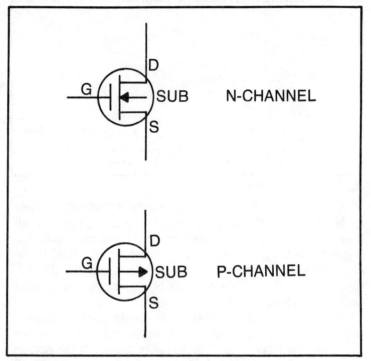

Fig. 5-5. Symbols for n-channel and p-channel devices.

MOSFETs, there is an inherent tendency for electrons to cluster close to the interface between the metal-oxide layer and the channel material. This will form a thin n-type region in the p-type substrate, located immediately beneath the insulating layer. The drain and source regions are n-type semiconductor material

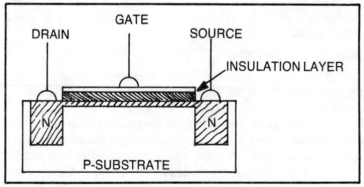

Fig. 5-6. Enhancement MOSFET with structure unbiased.

diffused into the p-type channel material. There is ordinarily no conduction between them, except for the tiny n-type region close to the gate structure. When the gate voltage is zero, the electrons of the n-channel are prevented from migrating out of the drain and source regions. But when a positive potential is applied (Fig. 5-7) to the gate, the electrons from these regions are attacted to the insulator-substrate interface, causing a conduction zone to appear between the drain and source. Current will flow under this circumstance. The conduction of the enhancement MOSFET is controlled by varying the gate voltage. This will increase or decrease the attraction of the drain and source electrons toward the gate region.

The circuit symbols for enhancement type MOSFETs are shown in Fig. 5-8. These symbols are the same as the depletion symbols, except that the channel bar is broken into sections to indicate the enhancement action taking place inside the transistor.

In both types of MOSFETs, the activity can be used as a switch. In both cases, one condition causes a low channel resistance, while another gate condition causes a high channel resistance. The on-resistance depends, in part, upon the signal voltage applied, but it is typically between 10 ohms and 2000 ohms. Most will have a value in the 100-ohm range. The variation of on-resistance is something of a problem and can lead to distortion of the output signal. But manufacturers have been able to limit this problem by correct FET switch topology. Designers can further limit the effects of the problem by using a load resistance that is high compared with the on-resistance of the channel. Load resistances between 10K ohms and 100K ohms are recommended.

The off-resistance of the channel is usually many megohms. The specification is usually given in terms of a *leakage current,* rather than a resistance. In most switches, the leakage current under the off condition, called the *effective off-resistance,* will be from 0.01 picoamperes to 100 picoamperes. Assuming a +10V applied signal, then, a "typical" 10 pA leakage current, the resistance would be:

$$R_{off} = \frac{10V}{1 \times 10^{-8}A} \qquad \textbf{Equation 5-1}$$

$$= 10^9 \text{ ohms} = 1000 \text{ megohms}$$

The on-resistance of the channel, on the other hand, drops to a much lower value, such as 100 ohms to 10,000 ohms. This fact,

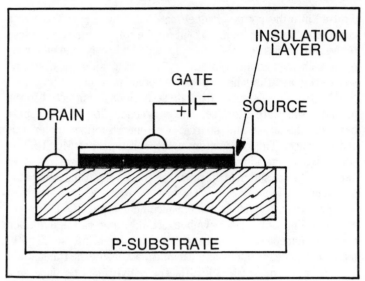

Fig. 5-7. Enhancement MOSFET with structure biased.

incidentally, makes the FET almost ideal for use as an analog electronic switch. A control voltage can be applied to the gate to

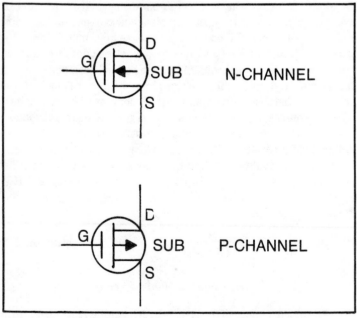

Fig. 5-8. MOSFET symbols.

obtain an expected switching action from the difference between on-resistance and off-resistance of the channel.

The primary use of the field-effect transistor is as an amplifier, or as the amplifying active element in an oscillator or active filter. The input impedance of the FET is typically very high. For some JFETs, on the low end of the price scale, this figure may be less than 1 megohm, but for most it is well over a megohm. For some MOSFET devices, the input impedance may exceed 1 terraohm (10^{12} ohms). In some circuit configurations, the high input impedance, coupled with a generally low output impedance, makes the device useful for service as an impedance converter. An amplifier for a signal source that has a high internal impedance can be made by using a field-effect transistor at the front end.

In the next chapter we will consider some aspects of FET circuit design and the types of circuits that might be selected. In Fig. 5-9, however, one elementary FET amplifier circuit is demonstrated. The circuit of Fig. 5-9 shows a common-source amplifier. The input impedance of this circuit is essentially the value of input resistance R1; the JFET input resistance is usually much higher than the value of this resistor. The input signal voltage is applied across the gate and source terminals of the JFET. The source terminal must be bypassed for AC by capacitor C3. The value of this capacitor must follow the one-tenth rule: The capacitive reactance of C3 must be one-tenth the value of R3.

The amplifier can operate as a voltage amplifier through a mechanism that is very similar to the scheme used to make a bipolar transistor operate as a voltage amplifier. Recall that in the case of the bipolar transistor (a current amplifier device) we "fooled" the circuit into voltage amplification by the use of the collector current causing a voltage drop across a collector resistor. In the case of the FET (a transconductance amplifier device), we "fool" the circuit into operating as a voltage amplifier by connecting a similar resistance (R2) in series with the drain. The channel of the FET operates as a resistance that varies with the applied gate voltage. When this voltage increases in the positive direction, the channel resistance drops, so the voltage at the drain is reduced (the percentage of the terminal voltage dropped across R2 increases). Similarly, when the drain resistance increases, the voltage at the drain increases. If the bias on the transistor is set at a point where the drain voltage is ½V+, the output voltage will swing positive and negative about this potential. The bias is set by the voltage drop across resistor R3, which is (analogous to cathode bias in

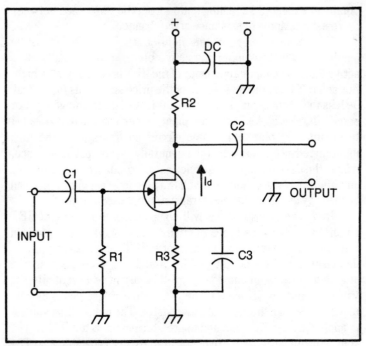

Fig. 5-9. JFET amplifier (common source).

vacuum tubes). This voltage is equal to the product of the source-drain current and the resistance of R3. We normally want the gate to be negative with respect to the source (this is an n-channel device), so making the source more positive than the gate serves the same purpose.

62

Chapter 6
Designing with FETs

There are several different types of field effect transistor. The two basic classes are the *junction field-effect transistor* (JFET) and the *metal oxide-semiconductor field-effect transistor* (MOSFET), also called the *insulated gate field-effect transistor* (IGFET). These different types were discussed in the Chapter 5.

THE TWO OPERATING MODES

There are two different operating modes for field-effect transistors: *depletion mode* and *enhancement mode*. It is generally true that many FETs will operate in either mode, depending upon the operating conditions. In the depletion mode, the channel current is maximum when the gate-source voltage is zero. In the enhancement mode, on the other hand, the channel current is minimum when the gate-source voltage is zero.

There are two regions of operation for a field-effect transistor. In the ohmic region, which exists for low drain-source voltages (V_{DS}), the channel current is proportional to the drain-source voltage. In the pinchoff region, which is also called the constant-current region, the current will not increase for further increases in drain-source voltage. It is this latter characteristic that allows the use of the junction field-effect transistor as a constant current source. In this application, the source and the gate are tied together, and the device operates as a diode with a constant current characteristic.

A field-effect transistor can be biased in any of several ways. We can use self-bias, external bias and certain combinations of self and external bias. Figure 6-1 shows two methods of FET biasing. Figure 6-1A shows an example of external bias. A reverse bias is applied to the gate of the field-effect transistor through resistor R1. The resistor is used to limit current and to provide a load for the input signal. It also allows the discharge of electrons that build up on the gate structure. The level of bias is set by varying E1.

An example of self-bias is shown in Fig. 6-1B. A JFET circuit is shown in Fig. 6-2. In this case, the voltage drop across source

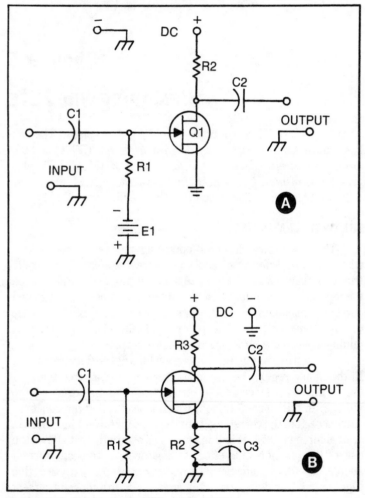

Fig. 6-1. (A) Battery bias for the FET and (B) self-bias.

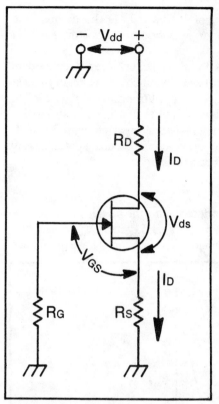

Fig. 6-2. JFET circuit.

resistor R2 is the bias voltage. The bias must make the gate more negative than the source. A negative voltage was applied to the gate in the previous case, but in this case the source is made more positive than the gate (conceptually the same thing). The value of the resistor must take into consideration the source-drain current of the FET and the desired level of bias. Ohm's law is applied. We know that the gate-source voltage is given by:

$$V_{gs} = -I_D \times R_s \qquad \textbf{Equation 6-1}$$

where V_{gs} is the gate-source voltage, I_D is the drain-source current, and R_s is the source resistor.

A power supply potential must be supplied to the FET (V_{dd}), but only a portion of the power supply voltage appears across the drain-source (V_{ds}). The voltage drops around the circuit are:

$$V_{dd} = I_D(R_d + R_s) + V_{ds} \qquad \textbf{Equation 6-2}$$

where V_{dd} is the power-supply voltage, I_D is the drain-source current, R_d is the drain resistor, R_s is the source resistor, and V_{ds} is the drain-source voltage.

We could select a value for the collector current (see the date sheet for the specific device) and a drain-source voltage. In most cases, the value of V_{ds} will be ½−V_{dd}. For example, if we applied 28V to the circuit (V_{dd}), and want 2 mA to flow in the drain circuit, we would select values for the resistors consistent with this goal. In general, we will make the source resistor approximately one-tenth to one-fifth of the drain resistor.

GRAPHICAL METHOD

One of the methods for determining the proper values for the source resistor is to use the graph of the I_d-versus-V_{gs} characteristic of the particular device under consideration. These graphs are

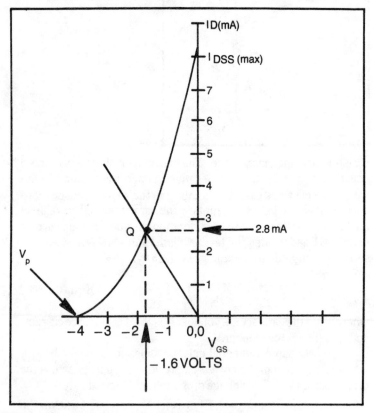

Fig. 6-3. Graphical solution for bias resistor.

printed in the manufacturer's data book and the specification sheets. An example is shown in Fig. 6-3. A line is drawn from the pinchoff voltage on the horizontal axis and the maximum drain current on the vertical axis. A proper drain current must be chosen. Then draw a line from the vertical axis at that point to the load line. The point at which this current line intersects the load line is directly over the proper gate-source voltage (V_{gs}). The slope of the line from the origin (0.0) to the *Q point* represents the resistance of the correct source resistor. In this case, the slope is:

$$S = \frac{(0.028A)}{(1.6V)} \qquad \textbf{Equation 6-3}$$

$$= 0.0175 \, \text{mhos}$$

$$R = 1/S = 57 \, \text{ohms}$$

Chapter 7
Bipolar Transistor
Power Amplifiers

The solid-state power amplifier solved the problem of size and bulk in amplifier designs. Vacuum-tube amplifiers required massive vacuum tubes, larger power transformers and, because of the high-voltage requirements, could not easily be operated in mobile or portable situations where only battery power was available. Older tube amplifiers required AC inverters, often vibrator power supplies (does that tell the author's age?) that produced a large amount of spurious output signal called *hash*. Transistors not only made the actual amplifier smaller, but could be operated directly from the 12 to 15 VDC power supplies available in automobiles. It is in the power amplifier that the differences between the vacuum-tube design and the more modern solid-state design are really pronounced. We also see a substantial improvement in the efficiency of the amplifier. The anode efficiency of the tube amplifier roughly approximates the collector efficiency of the transistor, but the transistor does not require filament (heater) power. The heater supply in the standard 25 to 50 watt audio power amplifier required another 25 watts just to heat up the vacuum tubes. Before considering some of the basic circuits that have become common, let's look at some transistor amplifier theory.

Figure 7-1 shows the current-versus-collector potential diagram for a bipolar power transistor. This graph shows the collector current caused by a given collector potential at various values of base current (I_{B1}, I_{B2}, and I_{B3}). The load line is drawn from a point

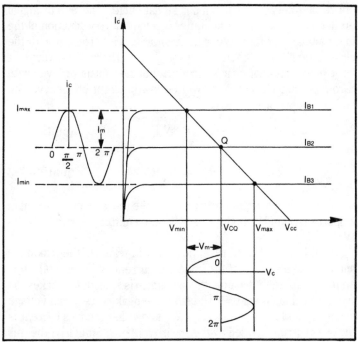

Fig. 7-1. Power transistor operating curves.

where the current is maximum (only $V_{ce(sat)}$ appears across the transistor collector-emitter terminals), to a point where the collector-emitter voltage is maximum ($I_c = 0$). An operating point will be established where, under quiescent conditions (no signal), the transistor is operated at point Q, which is approximately midway. The various values used in this graph are depicted in the schematic of Fig. 7-2. When an excitation signal, V_i, is applied to the base of the transistor, the base bias will vary the base current. This base current variation causes the operating point of the transistor to vary back and forth along the load line. The amplifier is theoretically linear, meaning there is no distortion, as long as the operating point is not shifted into one of the nonlinear regions of the family of curves.

CLASS-A AMPLIFIERS

The class-A amplifier is the only type of amplifier that is linear using only one transistor. Class-B can be linearized, but only at the expense of adding a second transistor to permit *push-pull* operation. In a class-A amplifier, the collector current flows over the entire 360 degrees of the input signal cycle. The output signal produced by the class-A amplifier is a faithful reproduction of the input signal, but it is larger in amplitude. The problem with the class-A amplifier is that it is very inefficient. There will always be a large power dissipation in the collector of the transistor, even with zero input signal, and there will be low inherent efficiency. The power dissipation is given by:

$$P = V_c I_c \qquad \text{Equation 7-1}$$

or

$$P = I_c^2 R_L \qquad \text{Equation 7-2}$$

where P is the power in watts (W), I_c is the rms collector current in amperes (A), and V_c is the rms collector-emitter potential in volts (V).

But how do we express the power in terms of the maximum and minimum swings of the collector current and C-E (collector-emitter) potential? Assume the waveform is sinusoidal (makes the arithmetic a sight easier!) and label the peak current and voltage swings I_m and V_m, respectively. We know that the rms current is the peak current divided by the square root of 2. Similarly, the rms collector-emitter voltage is the peak voltage divided by the square

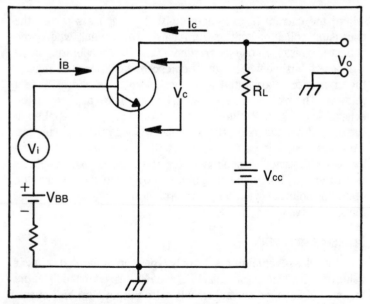

Fig. 7-2. Simple single-ended transistor amplifier.

root of 2. By taking the difference between I_m at the positive peak and I_m at the negative peak, expressions for power in terms of the voltage and current swings can be derived:

and

$$I_c = I_m/(2)^{1/2} = (I_{max} - I_{min})/(2)(2)^{1/2} \quad \textbf{Equation 7-3}$$

$$V_c = V_m/(2)^{1/2} = (V_{max} - V_{min})/(2)(2)^{1/2} \quad \textbf{Equation 7-4}$$

Assuming our symmetrical sine wave signal, we can write the power equation in the form:

$$P = V_m I_m/2 \qquad \textbf{Equation 7-5}$$

$$P = I_m^2 R_L/2 = V_m^2/(2)(R_L) \qquad \textbf{Equation 7-6}$$

which reduces to: $P = \dfrac{(V_{max} - V_{min}(I_{max} - I_{min})}{8} \qquad \textbf{Equation 7-7}$

Equation 7-7, then, is the expression of power in a class-A amplifier when the voltage and current extremes are known, as they can be from a load line graph. The values calculated from Equation 7-7, however, are valid only for a sine wave input signal.

Efficiency of the Class-A Amplifiers

The efficiency of any amplifier is the ratio of the signal power delivered to the load and the DC power consumed from the power supply. This does not include such "wasted" power as the filaments of a vacuum tube, but only the power used in the anode of the tube, or the collector of the transistor. We define the conversion efficiency factor η as

$$\eta = (P_L/P_{DC}) \times 100\% \qquad \textbf{Equation 7-8}$$

where η is the conversion efficiency in percent, P_L is the power delivered to the load, and P_{DC} is the DC power consumed from the power supply.

Equation 7-8 is a generalized case, useful for any amplifier. But, we are dealing with a specific amplifier, the bipolar transistor

71

power amplifier. In that case, we can substitute the values for the current and voltage that make up the power expressions demanded by Equation 7-8:

$$\eta = \; (\tfrac{1}{2} V_m I_m)(100\%)/(V_{cc} I_c) \qquad \textbf{Equation 7-9}$$

$$= \; (50)(V_m I_m)/(V_{cc} I_c) \text{ percent} \qquad \textbf{Equation 7-10}$$

This expression tells us something of the conversion efficiency, and from it we can make some inferences for the differences types of operation. In the small-signal case, for example, we know that the output power is very small, but we also know that the DC power consumed remains $V_{cc} I_c$, so the result is a very low conversion efficiency. We can also tell something about the large-signal behavior of the class-A amplifier. If the Q point is set exactly midway in the curve, then the collector current will swing from zero to the saturation current. In this case, I_m is equal to I_c. Similarly, the voltage between the collector and emitter will swing from a low of $V_{ce(sat)}$, which is essentially zero for our present purposes, to V_{cc} when the collector is totally cut off. In that case, V_m is equal to $\tfrac{1}{2} V_{cc}$. Substituting these values into Equation 7-10 will yield the maximum theoretical efficiency of the class-A amplifier under discussion:

$$\text{max} = \; \frac{(50)(V_m I_m)}{(V_{cc} I_c)} \; \text{percent} \qquad \textbf{Equation 7-11}$$

$$= \; \frac{(50)(\tfrac{1}{2} V_{cc})(I_c)}{(V_{cc} I_c)} \; \text{percent} \qquad \textbf{Equation 7-12}$$

$$= \; (50)(\tfrac{1}{2}) \text{ percent} \; = \; 25 \text{ percent} \qquad \textbf{Equation 7-13}$$

But just what does Equation 7-12 mean in practical terms. Let's consider the now standard 10 watts required of an ordinary automobile radio/stereo-tape unit. Many of these, especially those made as OEM (original equipment manufacturer) radios for the car makers, use class-A audio power amplifiers. For very 10 watts delivered to the load, 30 watts of DC power will be *wasted*. The total power requirement will be 40 watts. Note that the unused 30 watts doesn't just evaporate; it is used to heat the collector of the transistor. As a result, class-A power amplifiers require rather massive heat sinks for the transistors to dissipate the massive heat generated. In fact, this heat is so great that car radio manufacturers sometimes try to avoid placing the transistors on the front panel of the radio, lest they heat up the dashboard and elicit customer complaints!

Class-A Amplifier Circuits

The class-A amplifier is capable of delivering AC signal power with just one transistor. Figure 7-2 shows a popular class-A power amplifier used in the audio stage of many car radios. The output transistor is a pnp germanium device. Although these are now considered "old hat," in the light of new silicon devices, the pnp germanium transistor was the first power transistor commercially available in production quantities. In auto radios, there are only a couple of devices of any great popularity. The Motorola 2N176 and its successors, the Delco DS-501 and the Delco DS-503, were among the more popular. Delco replaced their versions, incidentally, with the DS-520 and DS-525 devices.

The output portion of this power amplifier consists of an impedance matching choke coil (L1) that is connected to the loudspeaker (LS). Some models, incidentally, connected the loudspeaker and the choke coil directly in parallel instead of using the tap arrangement shown. The collector of the PNP power transistor is kept close to ground for DC currents. The voltage drop across the coil, when the transistor is correctly biased and signal level is zero, will be around 1 to 1.5 volts. This particular circuit configuration, incidentally, is used in auto radios because the standard car electrical system is negative ground with the negative terminal of the battery is connected to chassis ground.

The signal is coupled to the base of the transistor through an interstage transformer (T1). This transformer is used to isolate the DC levels of the audio driver circuits from the base of the power transistor. An impedance transformer will match the relatively high impedance of the driver to the low impedance of the base circuit. This circuit is no longer used extensively, having been replaced by the direct-coupled circuits which will be described in a moment.

The bias for the power transistor is supplied by a resistor network (R2, R3 and R4). Note that the 0.2 to 0.3 VDC bias is a function of temperature, so resistor R3 is shunted by a high-value thermistor (R4). The value of the thermistor will change with changes in temperature, so it will help accommodate thermal tracking. Resistor R1 is used for two purposes: it is a very small value resistor (0.22 to 0.68 ohms); and it will produce a small amount of degenerative feedback, which helps reduce distortion, and acts as a fuse resistor. In fact, some manufacturers designate this resistor as a *fusistor*. It is designed to blow when the current level increases above a certain level, such as when the output current increases due to a shorted power transistor. Any 5W power resistor in the small value range *can* be a fusistor, especially if it is labeled for use in auto radios. Capacitor C1 is a large value electrolytic and is used to keep the cold end of the transformer (T1) secondary at a low impedance to ground for AC, while retaining a high DC-to-ground resistance.

A direct-coupled class-A power amplifier, also an auto radio output stage, is shown in Fig. 7-3. This circuit uses three transistors in a direct-coupled cascade arrangement. The output stage is a pnp germanium transistor that is very similar to the circuit in Fig. 7-4. The same tapped choke circuit is used to match the collector impedance to the loudspeaker impedance. Some of these circuits use an emitter fusistor as in Fig. 7-4; some do not.

Bias for transistor Q3 is set by the operation of transistor Q2. The collector current of transistor Q2 flows through resistors R1 and R2. It is the voltage drop across resistor R1 that sets the bias point of Q3. When Q2 is turned on harder, the collector current will increase, which increases the voltage drop across resistor R1. Increasing the voltage drop across R1 turns on the PNP output transistor harder.

Transistor Q2 is, in turn, controlled by transistor Q1. Both Q1 and Q2 are npn types, so they are turned on harder by increasing the base-emitter potential. Increasing the collector potential of Q1 will turn on Q2 harder. The only way to increase the collector voltage of Q1 is to turn off Q1. Decreasing the voltage applied to the base of transistor Q1 will tend to increase the collector voltage, thereby turning on Q2 and Q3. Some models will place a small variable resistor in the emitter circuit of Q1, which allows adjustment of the operating point of the entire stage. The usual procedure is to monitor the DC voltage at the collector of the output transistor (Q3), and then adjust the emitter resistor in the Q1 stage for the correct output voltage.

A direct-coupled class-A power amplifier using an npn silicon power transistor is shown in Fig. 7-5. The preamplifier and driver stages are similar to the previous example, but the output stage is completely different. The output transistor provides impedance

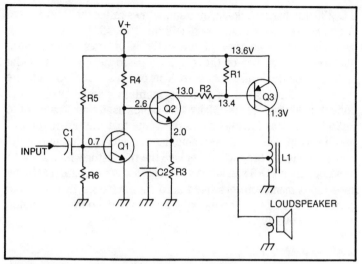

Fig. 7-3. Direct-coupled class-A audio power amplifier.

matching and degenerative feedback. Note that the secondary winding passes the DC current to the collector, while the primary is connected in parallel with the loudspeaker. The resistor in series with the transformer primary and the emitter of the power transistor is one of the low-value fusistor types.

The transistor generally used in circuits such as the one shown in Fig. 7-5 are the plastic case types. These devices use the

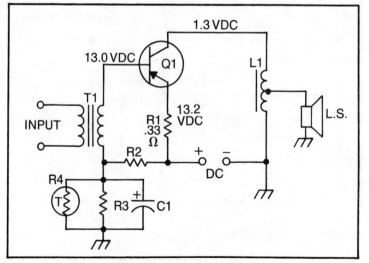

Fig. 7-4. Transformer coupled power amplifier (class-A).

75

TO-220 transistor package, also sometimes called the P-66 device. A typical transistor in this series is the 2N5249. One of the main problems with this type of transistor is an open emitter-base junction. The output of the stage drops to near zero, and the collector current is also near zero. This is in distinct contrast to the main problem found in pnp germanium power transistor: collector-emitter shorts. Many auto radios used this circuit. Auto radio technicians found early in the game that the transistor was unreliable in class-A power amplifier service at the auto radio power levels, so they would replace the transistor with a 2N3055 or some other TO-3 npn silicon device. It was fortunate that many auto radio manufacturers used auto radio chassis that still had either TO-3 or TO-66 holes drilled or molded, so the conversion was almost overly easy.

CLASS-B AMPLIFIERS

The class-B amplifier is defined as one in which the collector current flows over 180 degrees of the input cycle. The transfer characteristic of a typical class-B amplifier is shown in Fig. 7-6. The transistor receives no DC bias, so its operating point is set only by the input signal. This means that the collector current will flow on only one-half of the input signal excursion. If the transistor

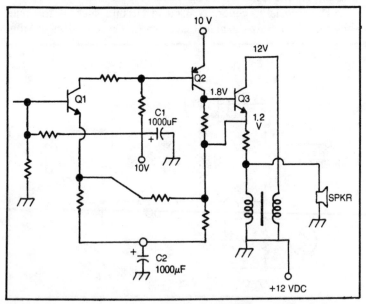

Fig. 7-5. Direct-coupled transistor class-A power amplifier using npn devices.

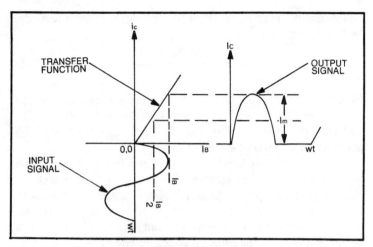

Fig. 7-6. Operating characteristic of the Class-B amplifier.

is npn, then the collector current flows only during the positive half of the input cycle. If the transistor is a pnp type, the collector current will flow on only the negative half-cycle of the input signal. The result is an output signal that closely resembles a half-wave rectified, amplified version of the input signal. But this presents a problem. An amplifier that produces only a half-wave output is *not* a linear amplifier. It cannot, therefore, be used as a linear audio power amplifier unless a second power transistor is used. We must contrive to process half of the signal in one transistor and the other half in the second power transistor.

Figure 7-7 shows a complementary symmetry push-pull power amplifier. The name, "push-pull," is derived from the fact that one transistor pushes while the other pulls. Note that two opposite polarity transistors are used: Q1 is an npn device and Q2 is a pnp device. The properties of the two transistors are exactly opposite. When the input signal is on the positive excursion of the cycle, npn transistor Q1 is turned on and pnp transistor Q2 is turned off. Similarly, on the negative half of the cyclic excursion, npn transistor Q1 is turned off and pnp transistor Q2 is turned on. This allows the outputs of the two devices to be summed, forming the full-wave output signal required for linear operation.

How much power can be obtained from a class-B power amplifier? What type of efficiency is to be expected? Well, let's see. We will first analyze each half of the push-pull pair in its turn. If the signal is sinusoidal, the DC current will be $I_{DC} = I_m/\pi$. This value is found by higher mathematics and is based upon the half-sinusoid. Given the value of DC current, the power level can be written as:

$$P = \frac{2I_m V_{cc}}{\pi}$$ **Equation 7-14**

Note that the "2" is because two transistors are feeding power to load resistor R_L. Equation 7-14 expresses the power. We can also find the efficiency and note that it will be approximately 78.5 percent for the perfect class-B power amplifier.

One advantage of the class-B amplifier is that the current drain goes to zero when the signal level is also zero. This makes the class-B push-pull power amplifier almost ideal for low-power portable projects and devices. Almost all transistor radios, for example, use class-B audio power amplifiers. The maximum power dissipation in the collector is expressed by:

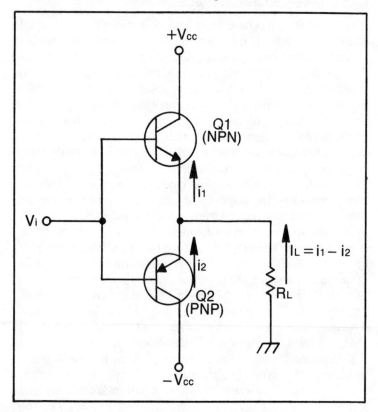

Fig. 7-7. Class-B push-pull amplifier using complementary symmetry.

$$P_{max(c)} = \frac{2V_{cc}^2}{\pi^2 R_L}$$ **Equation 7-15**

The maximum output power (when $V_m = V_{cc}$) is expressed by:

$$P_o = V_{cc}^2 / 2R_L$$ **Equation 7-16**

Combining,

$$P_{max(c)} = 4P_o / (\pi)^2$$ **Equation 7-17**

$$= 0.4P_o$$ **Equation 7-18**

CLASS-AB OPERATION AND CROSSOVER DISTORTION

Transistors are not totally linear devices when small signals are applied. There is a small necessary bias potential that must be overcome before the device becomes linear. This potential, called the *junction potential*, is 0.2 to 0.3 volts in germanium devices and 0.6 to 0.7 volts in silicon devices. Figure 7-8 shows the *actual* transfer property for a class-B push-pull power amplifier. The nonlinear region of the transfer curve produces the distortion shown in the output signal. This is called *crossover distortion* because it occurs when the signal is low level, or at the zero crossover points.

The crossover distortion problem can be overcome by using class-AB operation. This type of amplifier is defined as one in which the collector current passes for more than 180 degrees of the input cycle but less than the 360 degrees required of class-A. Class-AB, therefore, operates the transistor in a region that is between class-A and class-B.

The practical implementation of Class-AB operation is shown in Fig. 7-9. The transistors are given a small amount of forward bias by diodes D1 and D2. This will shove the transfer property farther up the curve into the linear region.

The output load—usually a loudspeaker—is isolated from the transistors with a DC blocking capacitor. The potential at the junction of the two transistors is approximately $\frac{1}{2}((V+) - (V-))$. This capacitor will have a value of 500 to 1000 μF in auto radios and up to 10,000 μF in high-power, high-fidelity amplifiers.

Class-AB Push-Pull Power Amplifier Circuits

Several different "standard" class-AB or class-B push-pull amplifier circuits have been developed over the years. Some of

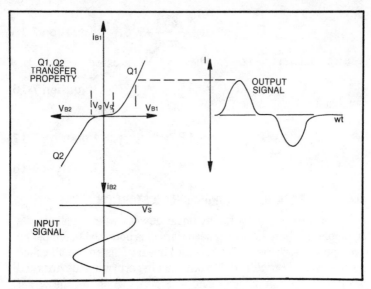

Fig. 7-8. Cross-over distortion.

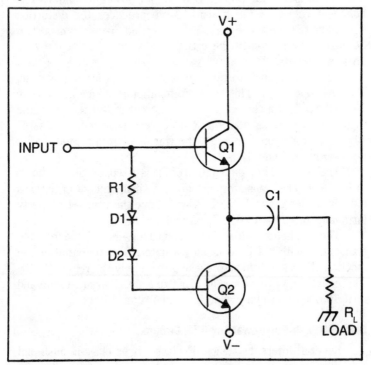

Fig. 7-9. Solution for cross-over distortion.

these circuits are merely transistorized versions of ideas that were common in vacuum-tube days, while others are completely different from anything that was used in tube days.

Figure 7-10 shows a basic push-pull power amplifier that is a holdover from older tube designs. This particular circuit was popular in transistor portable radios and a few communications applications, often as a modulator in an AM transmitter. The heart of the circuit is a pair of transformers, T1 and T2. Transformer T1 is the input transformer, also called the *interstage transformer* because it connects the power amplifier input to the output of the preamplifier or driver stages. The secondary of the input transformer is center tapped. If we take the center tap as the ground, or common, point, the signals at the two ends of the transformer will be out of phase with each other by 180 degrees. Capacitor C1 keeps the center tap of transformer T1 at AC ground potential, while allowing the DC potential required for bias to be above ground. The signals fed to the two transistor bases from the input transformer will be 180 degrees out of phase, so the criteria for push-pull operation will be satisfied.

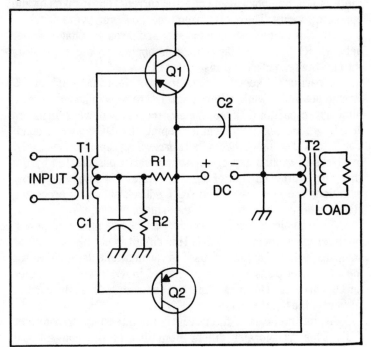

Fig. 7-10. Classical push-pull power amplifier circuit (class-AB).

The collectors of the two transistors are connected to the output transformer, T2. This transformer has a center-tapped primary, and the DC collector voltage is applied to the transistors through this center tap. Note that the collector voltage is negative with respect to ground, because the transistors used in the circuit are pnp types. This circuit is for negative-ground mobile operation. When there is no signal applied to the input transformer, the two transistors will be equally biased, so they will have equal collector currents. These currents will exactly cancel each other in the primary of T2, so the net output will be zero. When there is a signal applied to T1, however, one transistor will be cut off (during, say, the positive excursion of the input signal), and the other transistor will be turned on. This will unbalance the collector currents flowing in the primary of T2. The current balance is no longer equal, so there will be a net output that is proportional to the difference in collector currents in Q1/Q2. Similarly, on the negative excursion of the input signal, the opposite situation will occur. The transistor that had been cut off will then be turned on, and the transistor that had been on will be turned off. In that case, there is still an unbalanced collector current, but it takes on the opposite polarity. The output magnitude, however, is the same.

DC bias to the transistors is supplied through voltage divider network R1-R2. The bias voltage is applied to the transistors through the secondary of transformer T1.

Capacitor C1 keeps the center tap of the transformer at AC ground potential, while allowing the DC to remain above ground. Similarly, capacitor C2 keeps the emitters of the power transistors at AC ground potential, while keeping the DC above ground. Capacitor C2 is sometimes called a *decoupling* capacitor because it bypasses to ground any power-supply variations created by the changing emitter current of the transistors. These power-supply variations are seen as a valid AC feedback signal by preceding stages of the amplifier that share the same power supply.

An example of a split-secondary *totem pole* push-pull power amplifier is shown in Fig. 7-11. This circuit is not like any of the standard push-pull amplifiers used in vacuum-tube days. Note that the two power transistors are connected in *series* with each other for DC currents. This means that the collector of Q2 is the current source for the emitter of Q1.

Again, the heart of the circuit is the interstage transformer (T1). This transformer differs from that of the conventional push-pull power amplifier (Fig. 7-10) in that it has two secondary

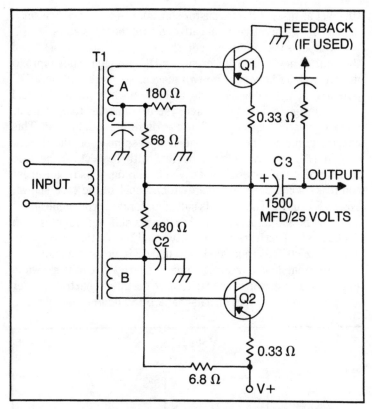

Fig. 7-11. Split-secondary totem pole push-pull power amplifier (class-AB).

windings. This arrangement is also sometimes called a *split secondary* transformer. Dots are used to indicate the phase sense of these secondary windings. The phase at the dotted ends will also be the same. To obtain push-pull operation, however, the two transistors must be driven 180 degrees out of phase with each other. This is done by feeding the respective transistor base terminals from opposite ends of their secondaries. The base of Q1 is driven directly from the dotted end of secondary *A*. Similarly, the base of transistor Q2 is driven from the undotted end of secondary *B*. The base terminals of the two transistors are biased from separate resistor voltage dividers, but the technique is the same as previously seen.

There are two low value resistors (0.33 ohms) used in this circuit. One 0.33-ohm resistor is placed in each transistor emitter circuit. These resistors serve two functions: they are fusistors (that act as fuses if the transistor shorts), and they supply a small amount of degenerative, or negative, bias to the transistors.

Output is taken at the junction between the collector of Q2 and the emitter of Q1. The power-supply DC voltage is split approximately equally between the two stages, so there will be a DC potential at this point equal to $(V+)/2$. Such a potential would cause damage to the loudspeaker system connected to the output, so a DC isolation capacitor (C3) is used in series with the load. This capacitor limits the low-frequency response of the power amplifier, so its value must be low with respect to the load resistance at the lowest frequency of operation. This requirement can make some very large capacitor values! In the case shown, 1500 μF is used. This value is sufficient for most communications and "low-fi" audio applications, but it is not sufficient for most hi-fi applications. There are high-fidelity amplifiers on the market with values to 25,000 μF, and 5000 to 10,000 μF is almost common.

An example of a transistor phase inverter circuit is shown in Fig. 7-12. This circuit is used to replace the interstage transformer and provides two output signals that are 180 degrees out of phase

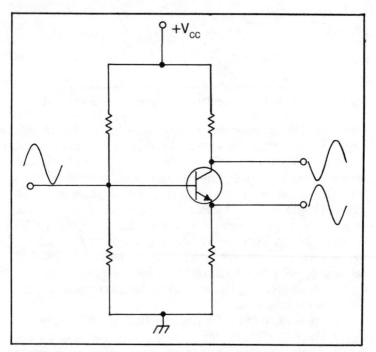

Fig. 7-12. Transistor phase inverter can replace the interstage transformer.

with each other. One signal is taken from the collector of the transistor, which means that it has a 180-degree phase shift with respect to the input signal. The other signal is taken from the emitter of the transistor (the stage operates as an emitter follower), so the signal is in phase with the input signal. This results in two signals that are 180 degrees out of phase with each other. It is necessary to adjust the values of the collector and emitter load resistors to make the amplitudes of the two signals equal.

In our initial discussion of class-B push-pull power amplifiers, we used two transistors of opposite polarity: an npn and pnp transistor. These two transistor types have opposite actions when driven by the input signal. A positive-going excursion of the input signal will turn the npn unit on and turn off the pnp unit. Similarly, on the negative half of the input signal, the npn will be turned off and the pnp will be turned on. The result will be a complete, almost linear reconstruction of the signal when the two emitter currents are summed together.

There is a problem in the design of this type of *complementary symmetry* amplifier. The two transistors are required to be *matched*—they must have the same properties except for polarity. This is not too much of a problem at low powers, but some early solid-state amplifier designers found only a few *complementary pairs* available from semiconductor manufacturers at the higher power levels. This situation resulted in the development of the *quasi-complementary symmetry* amplifier shown in Fig. 7-13. In this push-pull power amplifier, the output stage is a totem pole arrangement of two identical npn power transistors. The configuration is similar to that of the split-secondary totem pole amplifier shown in Fig. 7-11. The driver transistors, which operate at a lower power level, are in complementary symmetry. This is the circuit used in most hi-fi amplifiers on the market today and is even available in the form of integrated circuits and hybrid circuit modules.

MISCELLANEOUS TOPICS

The power sensitivity of an amplifier is increased markedly by using a *Darlington pair* (Fig. 7-14) as the output transistor. In some cases—not power amplifiers—the two transistors are identical. In most power amplifiers, however, the "output" transistor (Q2) will be a power unit, and the "input" transistor (Q1) will be a driver transistor. The current gain (*beta*) of the transistor pair is the product of the individual *betas*:

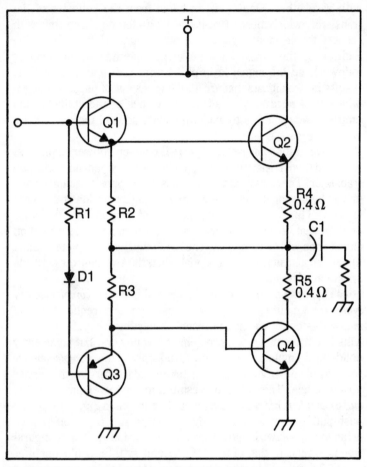

Fig. 7-13. Quasi-complementary symmetry push-pull power amplifier.

$$H_{fe} = H_{fe(Q1)} \times H_{fe(Q2)}$$

This greatly increases the sensitivity of the stage. In some hi-fi amplifiers, a Darlington pair is used in which both transistors are inside a common TO-3 transistor package. Motorola has made these, and many manufacturers of hi-fi equipment use them.

A new form of power amplifier has appeared on the market, and it is called *bridge audio.* Delco Electronics, the electronics manufacturing arm of General Motors Corp., offers an IC power amplifier that uses this principle (Delco DA-101). Figure 7-15A shows a common DC Wheatstone bridge. This circuit should be familiar to most electronics students and workers. The output

86

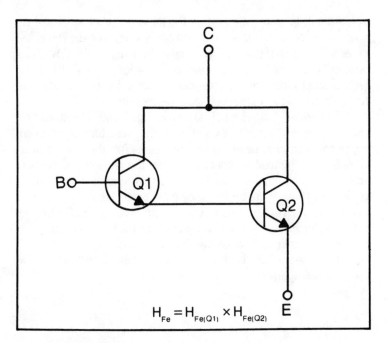

$$H_{Fe} = H_{Fe(Q1)} \times H_{Fe(Q2)}$$

Fig. 7-14. Darlington pair.

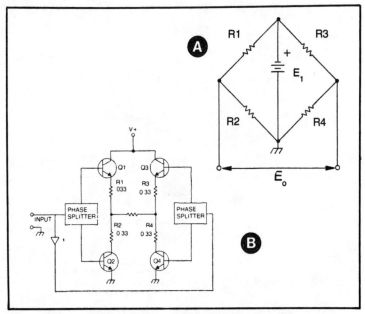

Fig. 7-15. Bridge audio with (A) Wheatstone bridge and (B) bridge audio circuit.

voltage will be zero when ratios R1/R2 and R3/R4 are equal to each other. Note that it is not strictly necessary that the resistors be equal, only that the ratios be equal. In many cases, however, simplicity is served best by making R1 = R2 = R3 = R4 = R. Unbalancing either voltage divider or both, in the Wheatstone bridge will cause an output voltage to be present.

The bridge audio circuit is shown in Fig. 7-15B. It is similar to the Wheatstone circuit, except that the resistors have been replaced with the collector-emitter paths of the four power transistors. The load is connected across the junctions of the two transistors, so E_o appears across the load. When a signal is applied to the two sides of the bridge, they will be unbalanced in opposite sense from each other, causing the voltage at one end of the load to rise and the voltage at the other end of the load to decrease. The result is a larger voltage swing than is possible with a standard totem pole amplifier. Each half of the circuit is, incidentally, a totem pole amplifier.

Chapter 8
Transistor RF Power Amplifiers

The vacuum tube held sway in the world of electronics for some 50 or more years. Beginning in the early- to mid- 1950s, however, the little device called the transistor (having been invented a few years earlier) began to replace the tube in electronic circuits. It was in 1962 that the first automobile radio for a major auto maker (General Motors) was fully transistorized, followed in 1963 by the other car makers. The transistor slowly eroded the position of the vacuum tube until today, when there are no longer any domestic tube makers producing receiving tubes, the transistor is all but totally involved in electronic circuitry. The only area where the vacuum tube is still king is in power amplifiers at radio frequencies. But don't think that transistors are not catching up! Until a few months before this was written, the tube was the *only* way to generate radio-frequency powers in the 500 watts and up range. But there is now a commercial all transistor rf power amplifier on the market that will produce a kilowatt (1000 watts). It is only a matter of time before we see the complete radio transmitter used in some medium-power AM band broadcast (BC) stations using all-transistor circuits. It is already possible to buy an AM BC transmitter with 250W driver stages. Amateurs can buy all-transistor radios with transistorized power amplifiers producing to 250 watts at 30 MHz. There are also 150W class VHF transmitters available (to 200 MHz) that are a dream for land mobile operation because they need no turn-up.

The first edition of the outline for this book did not consider rf power amplifiers, but it soon became apparent to the author that this subject could not be gracefully neglected in a book on amplifiers. The treatment is not extensive, but we will discuss rf amplifiers at least to some extent. You will gain the terminology and some basic understanding of the operation of these amplifiers.

SUITABLE AMPLIFIER CLASSES

The class-A amplifier is sometimes used a a power amplifier in audio circuits. Indeed, automobile radios and some communications receivers use nothing but class-A, single-ended power amplifiers. This is not usually done in the radio frequency range, however. In these ranges, we will find class-B amplifiers, class-AB amplifiers and class-C amplifiers. Class-C amplifiers are used in cases where the amplifier need not be linear, and they take advantage of the higher efficiency of the class-C amplifier circuit. Transmitters which might use the class-C amplifier include FM transmitters, CW transmitters (radiotelegraphy transmitters), and those AM transmitters in which the modulated stage is the final amplifier. If the modulation is impressed on a stage prior to the final amplifier, except in the case of FM transmitters, the following amplifiers (including the final) must be linear amplifiers. In those cases, the class-B or class-AB push-pull arrangement would be used. Some transistorized rf power amplifiers operate in class-B simply because it is very easy to build. A class-B transistor amplifier can be built with no bias resistor at all.

SIMPLE RF AMPLIFIERS

The simplest form of transistor power amplifier for radio frequencies is shown in Fig. 8-1. This circuit is often found in the buffer and driver stages of Citizens band transmitters. The input circuit consists of a parallel resonant tank circuit that is tuned to the frequency of operation. The coil in the tank circuit is tapped to match the impedance of the driving source. It is a fundamental property of circuits that transfer power that maximum power transfer occurs only when the load and source have the same impedance. Signals are passed to the base of the power transistor (Q1) through coil L2, which serves as a transformer secondary to L1.

The output circuit of Fig. 8-1 is also a parallel resonant tank circuit. In this case, the tap on the inductor is connected to the collector of the transistor, which is now the source of power. This

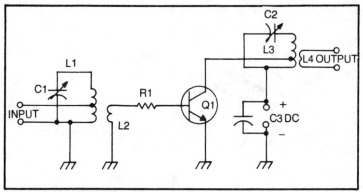

Fig. 8-1. Simple rf power amplifier.

particular circuit is called a series-fed amplifier because the DC current passes through the inductor of the tank circuit before it reaches the collector of the transistor. A parallel-fed, or shunt-fed, amplifier would not have the collector current pass through the tank circuit coil. Output signal to the load is taken from a low-impedance transformer secondary. A power-supply bypass capacitor (C3) is connected across the power-supply terminals. This capacitor insures that the power-supply impedance is low to the resonant frequency of the tank circuit. The capacitor also provides decoupling to other stages that share the same power supply.

If you want the amplifier to operate class-B, delete resistor Rl. In that case, the transistor is zero-biased, so it will operate only on the positive half-cycles of the input signal. When the series resistor (R1) is in the circuit, the stage will operate class-C. This is because a small current through the resistor will set up a DC bias voltage drop across the resistor. It is this voltage drop, which reverse biases the transistor, that makes the stage operate class-C.

RF POWER TRANSISTORS

Some of the first commercially available rf transistors were intended for use in CB transmitters. These devices would produce several watts of rf energy at 27 MHz and were terribly expensive; $15 was not an uncommon price. To make matters worse, those transistors could not accept a high VSWR (voltage standing-wave ratio) in the transmitter/antenna circuit. The transistors would pop if an anomaly developed in the antenna circuit. Many CB operators lost their power transistors because they operated the transmitter

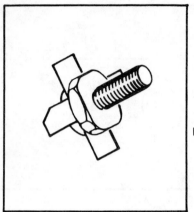

Fig. 8-2. Typical rf power transistor.

with the antenna disconnected (a \$35 mistake). Most modern transistors will sustain a higher VSWR than the earlier models, will generate far greater power levels and will operate at much higher frequencies. Figure 8-2 shows a typical rf transistor package. The threaded stud is used to mount the transistor to the heat sink and is usually connected to either the collector of the emitter terminal of the transistor. The wings, which make this device look like a space satellite, are the collector, base and emitter terminals of the transistor. The wide, flat shape of these wings is intended to keep the lead inductance low. It is the lead inductance that prevents many transistors from operating at high frequencies.

TYPICAL 30-MHz CIRCUIT

Some solid-state rf power amplifiers are tuned, while others are broad banded. Figure 8-3 shows a circuit that is tuned to

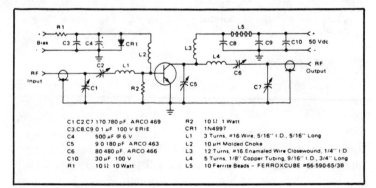

C1 C2 C7 170 780 pF ARCO 469	R2 10 Ω 1 Watt
C3,C8,C9 0 1 μF 100 V ERIE	CR1 1N4997
C4 500 μF @ 6 V	L1 3 Turns, #16 Wire, 5/16" I.D., 5/16" Long
C5 9 0 180 pF ARCO 463	L2 10 μH Molded Choke
C6 80 480 pF ARCO 466	L3 12 Turns, #16 Enamaled Wire Closewound, 1/4" I D
C10 30 μF 100 V	L4 5 Turns, 1/8" Copper Tubing, 9/16" I.D., 3/4" Long
R1 10 Ω 10 Watt	L5 10 Ferrite Beads – FERROXCUBE #56-590-65/3B

Fig. 8-3. A 30-MHz power amplifier.

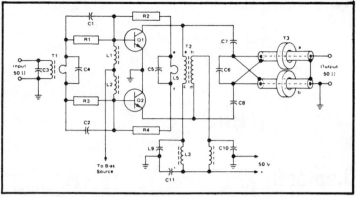

Fig. 8-4. A 300-watt HF power amplifier.

operate at approximately 30 MHz. The 50-ohm input impedance (standard for rf power amplifiers) is reduced to approximately 5 ohms at the base of the power transistor by an LC network consisting of C1, C2 and L1. DC bias which would be needed if the circuit is to be a linear power amplifier to amplify AM of SSB signals. Then forward bias is needed. The collector impedance matching network consists of C5, C6, C7 and L4. The output impedance is 50 ohms.

The collector current passes to the collector of the power transistor through rf choke L3 and another rf choke L5. Together with capacitor C8, this circuit forms a low-pass filter to decouple the collector signal to ground before it reaches the power supply. Rf choke L5 is actually a series of ferrite beads instead of an actual coil. Ferrite donuts are slipped over the wire that carries the collector current to the transistor. Such beads are known to possess inductance and are used extensively in HF and VHF transmitters.

Figure 8-4 shows a method for obtaining higher power levels. There are two basic ways to increase the power level in an amplifier: parallel two or more devices, or connect them in push-pull. This circuit connects the transistors in push-pull. Note the output circuit. The signals from the two halves of a push-pull amplifier must be summed in the output circuit, and this type of circuit offers that service using a transmission line transformer method. Each half of the circuit operates in a manner similar to that of the previous example.

Chapter 9
Introduction to
Operational Amplifiers

The operational amplifier has, perhaps, revolutionized electronic design second only to digital electronics. The operational amplifier was originally invented for a relatively narrow application, but has become the standard building block of analog circuit designers. It is a lot easier to apply than either the bipolar or field-effect transistor, and is so simple that even relative novices in the arcane art of circuit design can become proficient. It requires only a little skill to design with operational amplifiers, so the low cost of the integrated circuit operational amplifier device makes their use almost assumed.

The original intent of the designers of the operational amplifier was to perform mathematical *operations* in analog computers. These instruments have been eclipsed by the digital variety in recent years, but once were the only computers available. The analog computer, incidentally, is still the only way to solve certain differential equations. A large number of electronic instruments used in science, engineering and medicine are actually analog computers with specialized programing and control circuits. While digital circuits, especially microprocessors, have made inroads into this area of design, the analog computer-like circuitry of electronic instruments is still very much in evidence.

OP AMP BASICS

But what is this device known as the operational amplifier? How does it differ from other types of amplifiers? Why is it so useful and easy to apply? All operational amplifiers are intended to have an external feedback network, and it is the feedback network that sets the properties of the overall amplifier. Let us consider a basic thought experiment. Figure 9-1 shows a feedback amplifier system. The amplifier portion of this circuit has a gain of A_{vol}, while the gain (transfer function) of the feedback network is represented by Greek letter *beta* (β). We know from elementary feedback theory that the overall gain of the system will be:

$$A_f = \frac{A_{vol}}{1 + A_{vol}\beta} \qquad \textbf{Equation 9-1}$$

where A_f is the gain of the system when the feedback loop is connected, A_{vol} is the voltage gain of the circuit when the feedback loop is open-circuited (the open-loop gain), and β is the transfer function of the feedback network.

In ordinary amplifiers, the value of the open-loop gain will usually be low, and the transfer function of the feedback network will deliver a small fraction of the output voltage to the input. The study of feedback amplifiers is, incidentally, a study in its own right.

But we are considering the operational amplifier, not ordinary feedback amplifiers. The operational amplifier, by definition, has an infinite open-loop gain. Under this condition assume that the value of Equation 9-1 will reduce to $1/\beta$, meaning that the transfer function of the overall system is determined solely by the transfer function of the feedback network (β).

The ideal operational amplifier is defined as having the following six properties.

☐ Infinite open-loop gain ($A_{vol} = \infty$)
☐ Infinite input impedance ($Z_{in} = \infty$)
☐ Zero output impedance ($Z_o = 0$)
☐ Infinite open-loop bandwidth
☐ Noise figure $= 0$
☐ Inputs stick together (of which, more in a moment)

We have discussed the meaning of the first of these requirements in the preceding section. When the gain of the operational amplifier is infinite, the overall gain of the circuit *with* feedback (A_f) is controlled solely by the transfer function of the feedback network. Real operational amplifiers do not have infinite open-loop gain, but the approximations are stunning. Typical "garbage-grade" cheapies will exhibit open-loop gain figures of 20,000 to

100,000, while some of the premium-grade devices have an open-loop gain over 1,000,000! For most applications, these figures are near enough to infinite for the operational amplifier-like behavior to be apparent. Even some low-cost, modern operational amplifiers are capable of providing open-loop gain figures that are closer to what was considered "premium" only a few short years ago.

The second requirement is infinite input impedance. The value of this is that the output voltage is in no way a function of input loading of the driving source. In most amplifiers, the actual input voltage is a fraction (sometimes a large fraction, but still a fraction) of the source voltage. In an operational amplifier, the loading of the source by the amplifier input is eliminated. There is one result of this requirement that must be remembered and will show up later in our discussion of the op amp transfer equation derivations: *The inputs of an operational amplifier will neither sink nor source current!* How can we make this statement? Well, let's get back to basics. What is input impedance? We know that the input impedance of an amplifier is $Z_{in} = V_i/I_i$, where Z_{in} is the input impedance in ohms, V_i is the input voltage, and I_i is the input current. Solving for I_i then gives us $I_i = V_i/Z_{in}$. Because Z_{in} is infinite, the input current goes to zero.

Again, real operational amplifiers do not approach the ideal. Once again, the approximations are exciting. The most common operational amplifiers will offer input impedances in the 500,000 ohms range, while some of the newer MOSFET input operational amplifiers will offer input impedances of 1.5 terraohms (1.5×10^{12})! If this isn't a good approximation of infinite, as far as ordinary circuit impedances are concerned, then nothing is.

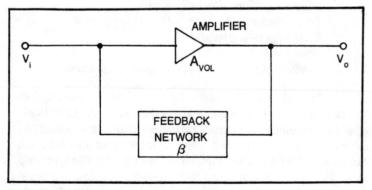

Fig. 9-1. Block diagram of feedback amplifier.

The third requirement is that the output impedance must be zero. This is a requirement of any ideal voltage amplifier and is needed because we want to eliminate the voltage divider effect of typical amplifier output circuits. Assume that the output impedance of an amplifier is Z_o and the impedance of the load is Z_L. If V_o is the output voltage with the load disconnected, the actual output voltage (V'_o) when the load is connected will be:

$$V'_o = \frac{V_o Z_L}{Z_o + Z_L} \qquad \text{Equation 9-2}$$

You may recognize this equation as the standard voltage divider equation from elementary electronic theory. The output impedance of the amplifier forms, with the load impedance, a voltage divider that produces a reduced output potential. If Z_o is set equal to zero, the equation reduces to:

$$V' = \frac{V_o Z_L}{Z_L} \qquad \text{Equation 9-3}$$

$$V'_o = V_o \qquad \text{Equation 9-4}$$

In the ideal operational amplifier, we want to make the transfer function dependent solely upon the transfer function of the feedback network. If there is an appreciable output resistance, though, a component must be added to the equation that will account for the phenomenon described by Equation 9-2. The ideal operational amplifier has an output impedance of zero, but the real operational amplifier will have some nonzero output impedance. Once again, the approximations are exciting. Real operational amplifiers are available with output impedances as low as 1 ohm, with almost all being under 100 ohms. It is, perhaps, safe to say that all operational amplifiers commonly used have output impedances in the 20 to 80 ohms region.

A standard rule of thumb used extensively in engineering is to use a load impedance that is at least 10 times the source impedance. In an amplifier circuit, output impedance Z_o is the source impedance, so load resistances as low as 1000 ohms (worst case is $Z_o = 100$ ohms) will provide ± 10 percent accuracy of the transfer function. As a practical example, let us use an operational amplifier that has an output impedance of 20 ohms. The 10-times rule means that the load must be 200 ohms for a 10 percent gain error, or 2000 ohms for a 1 percent error. If we increase the load impedance to only 20,000 (a typical value, actually), the gain error is approximately 0.1 percent.

The ideal operational amplifier will have an infinite open-loop bandwidth (condition No. 4) and zero noise contribution (condition No. 5). Some operational amplifiers have gain-bandwidth products of 10 to 20 MHz, which closely approximates "infinity" for DC and low-frequency AC signals. But note that wide bandwidth also produces many other problems, which are difficult to handle. The layout of the circuit will become extremely critical, and oscillations result if some of the layout rules are violated. This makes the amplifier unstable, and a pain in the neck to use. Many modern operational amplifiers are said to be compensated, meaning that the bandwidth is purposely limited in order to make the circuit unconditionally stable. The 741 is probably the most famous example of the compensated operational amplifier. The gain-bandwidth product of this popular device is only a few kilohertz. This limits its usefulness as a sideband amplifier but in no way affects the usefulness of the device at low frequencies and DC. Because most circuit applications require bandwidths less than 10 kHz, the 741 is often the device of choice. Its principal appeal, incidentally, is that it is dirt cheap.

Some operational amplifiers are available that do not have any internal frequency compensation. These will be called *uncompensated*, or *externally compensated* devices. If the specifications for any particular device seem silent on this issuse (they often are!), look at the diagram for the package pinouts. If there is a compensation terminal, or a terminal marked "lag" or "lead," it is an uncompensated type.

The noise contribution of the operational amplifier is ideally zero. We know that it is impossible to make any amplifier with a zero internal noise figure, but some operational amplifiers are good approximations of the ideal. But note that most common, low-cost operational amplifiers are anything but low noise. Some of these devices are terrible in this respect, and this is the one area where most operational amplifiers do not have a good approximation of the ideal op amp of textbook discussions. Low-noise operational amplifiers on the market, however, but these are premium devices and carry price tags that reflect that fact. Most operational amplifiers cannot be used in circuits with low-level signals, because of their own internal noise problems. But the nature of noise in amplifier circuits is such that we can often approximate the low-noise, high-gain amplifiers needed in these cases by using a low-noise operational amplifier in the first stage of a casade chain and then regular operational amplifiers in the following stages. This will preserve the low-noise capability needed and still provide us with a low-cost amplifier, albeit not as low as when all cheapies are used.

The sixth rule for operational amplifiers is that all inputs stick together. This needs a little explanation. The ideal operational amplifier needs only one input, but most real operational amplifiers have *two* inputs. Because of this fact, most people think that all operational amplifiers must have two inputs.

The two inputs of the typical real operational amplifier form a *differential* input-pair. There are two types of input: *inverting* and *noninverting*. These inputs will have equal but opposite effects on the output voltage (both look into the same gain). But their phases are opposite. The inverting (−) input produce an output that is 180 degrees out of phase with the input signal; i.e., it is an inverter. The noninverting (+) input produces an output that is in phase with the input signal. A very useful result of this differential pair of inputs is that the output voltage will be proportional to the difference between the voltages applied to the two inputs. This theme will be developed further in Chapter 12 when we discuss the vaious differential amplifier forms.

What do you suppose happens in the two-input operational amplifier when equal voltages are applied to the two inputs—the same voltage with the same polarity? This is simulated by tying the two inputs together and then applying an input signal voltage. Because both inputs look into the same close-loop gain, they produce equal but opposite polarity effect on the output voltages. In this case, the contributions from the two inputs algebraically add, causing total cancellation of the output voltage. The output voltage, then, will be zero when the inputs are at the same input voltage.

This phenomenon is used to good effect in differential amplifiers and will tend to cancel certain types of interference in real equipment. Medical amplifiers designed to pick up the electrocardiograph (ECG), are differential, so will cancel the 60-Hz hum signal from the local power lines. The human ECG signal is on the order of 1 millivolt, while the interference can be several volts! You can see why the differential nature of the operational amplifier inputs are so useful.

But what do we mean by "stick together"? The standard operational amplifier inputs will tend to track each other such that we must treat them as one entity mathematically. If, for example, we apply a voltage at one input, we must treat the other input (mathematically) *as if it were at the same voltage*! This is not merely some theoretical device only useful in textbooks, but a real phenomenon. If we were to build an operational amplifier circuit and apply , say, +3 volts to the inverting input, a voltmeter would measure +3 volts at the noninverting input as well. That voltage is real and can be measured. Not to labor this point too much, but it is necessary because of the material that will be presented in the next few chapters.

OP AMP CIRCUITS

The operational amplifier must provide a very high gain (A_{vol} greater than 20,000 and up to several million) and produce little or no input bias current. Figure 9-2 shows the basic differential input

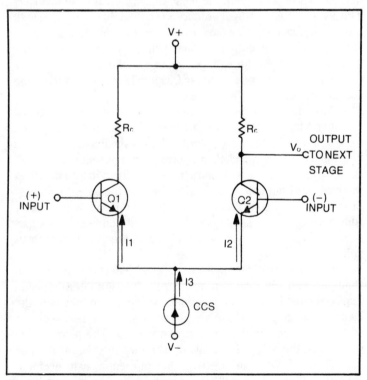

Fig. 9-2. Differential transistor amplifier.

circuit used on many common operational amplifiers. Transistors Q1 and Q2 form a *differential pair*, and the bases of these transistors form the op amp input terminals.

Consider the collector of transistor Q2 to be the output terminal. In real operational amplifiers, this terminal will actually be used to drive the stages following the input amplifier, so is a good starting point. We will consider what happens to the voltage at this point (V_o) in response to signals applied to the bases of Q1 and Q2.

The emitters of Q1 and Q2 are tied together and are fed by a constant current source (CCS). In the actual op amp, the CCS will be another transistor or transistor-pair, but the result is the same: I3 is constant regardless of changes elsewhere in the circuit.

We know from Kirchhoff's current law (KCL) that the currents in the circuit of Fig. 9-2 have the following relationship:

$$I_3 = I_1 + I_2 \qquad \textbf{Equation 9-5}$$

Equation 9-5 tells us that any change in one of the transistor emitter currents (I1 and I2) must be reflected as a change in the other emitter current, because I3 will always be *constant*. If, for example, current I1 increases, current I2 must decrease. Similarly, when current I1 decreases, I2 must increase.

The output voltage is a fraction of the supply voltage spread $(V+) - (V-)$. It is the voltage drop across Q3 collector resistor R_c that determines the value of V_o. If the collector current of Q2 increases, the voltage drop across R_c increases (Ohm's law), leaving less of the supply voltage for V_o; V_o, therefore, decreases. Similarly, a decrease in the Q2 collector current decreases the voltage drop across R_c, leaving a larger percentage of the supply voltage for V_o. The output voltage may then be affected by varying the collector current. For sake of simplicity, we will consider the collector currents of Q1 and Q2 to be equal to the emitter currents. In real transistors this is not too great a departure from common sense because the base current is only 1 to 5 percent of the collector current, making the emitter current nearly equal to the collector current. In the hypothetical operational amplifier, when these base currents are zero, this assumption is even more justified.

Consider the circuit of Fig. 9-2, remembering the foregoing discussions. We will designate the base of transistor Q1 as the noninverting input and the base of transistor Q2 as the inverting input. First, deal with the inverting input. Suppose that a sine wave is applied to the base of Q2. What happens? On the *positive-going* half of the sine wave, the *collector current of Q2 will increase*, while on the *negative-going* half of the sine wave the *collector current will decrease*. How does this affect the output voltage, V_o? When the collector current increases (positive half of the sine wave), V_o will decrease; note the phase inversion. Similarly, when the collector current in Q2 decreases (negative half of the input sine wave applied to the base of Q2), the value of V_o will increase. Again, note the phase inversion. We may conclude, then, that our assumption of the base of transistor Q2 as the inverting input was a correct selection.

Now turn to the noninverting input, the base of transistor Q1. When both inputs are at the same voltage, the collector currents of Q1 and Q2 will be equal. The value of the output voltage, V_o, will be at some quiescent voltage (zero volts when V− and V+ are equal). Again assume that a sine wave is applied to the input, in this case the noninverting input. When the sine wave is positive-going, the collector current of Q1 (I1) will increase. When the sine wave applied to the noninverting input is negative-going, collector current I1 will decrease. Recall that I3 is a constant and that it is equal to the sum of I1 and I2. When I1 increases on the positive-going portion of the input sine wave, I2 must decrease in order to maintain the constancy of I3. This means that output voltage, V_o, will increase. Both the input voltage and the output voltage, then, are increasing; there is no phase inversion. On the negative half of the input sine wave, current I1 will decrease, causing I2 to increase. This will force output voltage V_o to decrease also. Again, there is no phase inversion.

A real operational amplifier will have many additional stages after the input preamplifier. These stages are used to build up the voltage gain of the device and are easy to come by with IC technology. The last stage in the op amp will be a power amplifier stage, although the *power* developed will often only be a few milliwatts to a few dozen milliwatts (op amps are voltage amplifiers, not power amplifiers, unless otherwise stated in a particular specifications sheets).

Figure 9-3 shows the basic circuit symbol for the operational amplifier, whether IC or discreet technologies are used in the fabrication. The symbol is a triangle with the output taken at one of the apexes. The inverting (−) and noninverting (+) inputs are along the side of the triangle opposite the apex chosen as the output terminal. It is normal to label the inverting and noninverting inputs with the − and + signs, respectively.

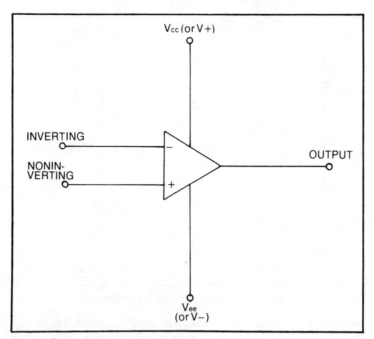

Fig. 9-3. Symbol for an operational amplifier.

The power supply terminals are also shown in Fig. 9-3. There are two terminals, one each for positive and negative power supplies. It is common practice to delete these terminals in actual circuit schematics to reduce the clutter of the drawing. These terminals are, nonetheless, connected, or the op amp wouldn't work! Keep in mind, however, that there may be no power-supply terminals shown on many diagrams.

The typical op amp power-supply configuration is shown in Fig. 9-4. The true operational amplifier was required to perform mathematical operations in all four quadrants, so it had to be capable of producing either positive or negative output potentials. In most cases, the positive voltage (positive with respect to ground) is called V_{cc} (after transistor terminology) or V+. The negative-to-ground voltage will be designated V_{cc} or V−. These voltages are represented by batteries B1 and B2, respectively, in Fig. 9-4.

Note that there is no ground terminal on the operational amplifier symbol of Fig. 9-3. This is a little difficult for some people to accept but is the common case. The input and output potentials of the operational amplifier are ground referenced, however,

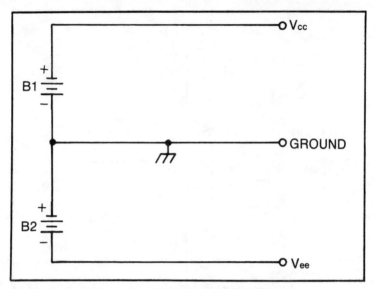

Fig. 9-4. Power-supply configuration for operational amplifiers.

meaning that the ground connection must be provided by the power-supply configuration. I once knew a famous editor of an electronics magazine who was extremely knowledgeable about all phases of electronics. When he sent me the proofs to an article on operational amplifiers, however, I found that he had drawn in ground terminals on the op amps, in addition to the V+ and V− terminals. I had to call him on the telephone and give him the reasons why this was improper. This veteran of many years in electronics did not know that op amps have no ground terminal, so don't be embarrassed if you make the same mistake.

OP AMP PACKAGES

Operational amplifiers are available in all of the various integrated circuit packages. The most standard is probably the 8-pin miniDIP, while some are also available in each of the others. The 8-pin metal can was one of the earliest types and is still seen when one is buying a premium-grade operational amplifier. It is worth noting that the pinouts for the miniDIP are usually the same as for the 8-pin metal can.

Some high-grade operational amplifiers are not integrated circuits at all. Many of the best operational amplifiers are hybrids, meaning a combination of discrete and integrated technology. This approach allows trimming of the circuit parameters to provide more nearly ideal performance of the device.

Chapter 10
Inverting Followers

One of the most common amplifier circuits using the operational amplifier device is the *inverting follower*. The input signal will be applied to the inverting input of the operational amplifier device. In most cases, the noninverting input will be grounded.

The circuit for a typical inverting follower is shown in Fig. 10-1. There are two resistors, one in the feedback loop and the other in the input circuit. In this case, the noninverting input is grounded. We will analyze this circuit in terms of Kirchhoff's current law and Ohm's law.

VIRTUAL GROUND

A concept that tends to both many new students of operational amplifier circuitry is the idea of a virtual ground. This concept is derived from the sixth condition for an ideal operational amplifier: the inputs stick together. Consider Fig. 10-1. The noninverting input is grounded, which is the same as saying that a potential of zero volts is applied to the noninverting input. Recall from Chapter 9 that applying a potential (in this case, 0 volts) to one input forces us to treat the other input *as if it were at the same potential*. In the case shown, we must treat the inverting input terminal of the operational amplifier as if it were also grounded, hence the term, *virtual ground*. If we were to take a voltmeter and measure the potential on the inverting input of the operational amplifier in Fig. 10-1, we would indeed find zero volts.

ANALYSIS OF THE INVERTING FOLLOWER

The point labeled "A" in Fig. 10-1 is known as the *summing junction*, because all currents will sum at this junction. Input current I1 is due to the action of input voltage E_{in} on the input resistor. Feedback current I2 is due to the action of output voltage E_{out} on the feedback resistor. There may also be a bias current I_o from inside of the operational amplifier, but this current will be zero in ideal op amps. In real op amps, the current may be very low, with picoamperes being claimed in some premium units. For the present discussion, consider I_o to be equal to zero.

We know the values of the two active currents in Fig. 10-1. The input current is set by the input voltage and the value of the input resistor because point A is at ground potential (virtual ground, remember?):

$$I1 = E_{in} / R_{in} \qquad \textbf{Equation 10-1}$$

The feedback current is similarly derived:

$$I2 = E_{out} / R_f \qquad \textbf{Equation 10-2}$$

By Kirchhoff's current law, we know that the following is true:

$$I1 + I2 = 0 \qquad \textbf{Equation 10-3}$$

$$I2 = -I1 \qquad \textbf{Equation 10-4}$$

Now substitute into Equation 10-4 the values for I1 and I2 as defined in Equations 10-1 and 10-2:

$$\frac{E_{out}}{R_f} = \frac{-E_{in}}{R_{in}} \qquad \textbf{Equation 10-5}$$

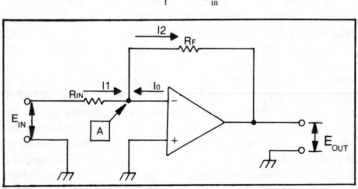

Fig. 10-1. Inverting follower amplifier.

Now evaluate Equation 10-5 in terms of E_{out}, R_f and R_{in}:

$$E_{out} = \frac{-E_{in} R_f}{R_{in}} \qquad \textbf{Equation 10-6}$$

Equation 10-6 defines the output voltage in terms of the input voltage and the feedback network. The minus sign used in Equation 10-6 denotes that the output voltage is 180 degrees out of phase with the input voltage, the normal situation for inverter circuits.

The term, R_f/R_{in}, is known as the gain of the circuit and reflects the gain of the circuit with *feedback*. The exciting thing about operational amplifiers is that, for frequencies less than about 100 kHz, we can use the ratio of the feedback and input resistances to set the circuit gain! For example, if we want to build an amplifier with a voltage gain of 100, we would make sure that the ratio of the feedback to input resistors is exactly 100. We sometimes see Equation 10-6 written in the form:

$$E_{out} = -A_v E_{in} \qquad \textbf{Equation 10-7}$$

where E_{out} is the output voltage, E_{in} is the input voltage, and A_v is the factor R_f/R_{in}.

□ **Example:**
It is instructive to look at a design example when dealing with electronic circuits. Operational amplifier circuits are no different, despite their utter simplicity. Figure 10-2 shows a gain of 50 inverting amplifier based on an operational amplifier. How did we arrive at the resistor values? We know that they must have a ratio of 50:1 to produce the required gain of 50, but why these particular values?

Part of the answer is that it is almost arbitrary—not quite—but a wide latitude is allowed in the selection of values. Sometimes, the exact selection might be made on the basis of available stocks. Other times, you might have to make the selection based upon the source impedance. We follow a rule that says that the input impedance of the amplifier be *at least* 10 times the source impedance. If we want even better accuracy, we would have to have an input impedance of 100 times the source impedance. This will result in a gain error of 1 percent. Assume that the source used to drive this amplifier has an output impedance of 100 ohms. Using the times-10 rule, you would want an input

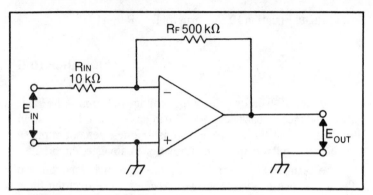

Fig. 10-2. Gain-of-50 amplifier.

resistor of at least 1000 ohms. With the times-100 rule, you would want 10,000 ohms. We have selected the times-100 rule, meaning that R_{in} is set to 10K ohms. This means that the feedback resistor must have a value of:

$$R_f = 50(R_{in}) \qquad \textbf{Equation 10-8}$$

$$= (50)(10,000 \text{ ohms})$$

$$= 500,000 \text{ ohms}$$

In practical circuits, where standard value components must be used, it is not always possible to obtain the precise value resistors needed to obtain the specified gain. In other cases, we will be able to obtain the correct value resistor only to find a small gain error. This error is due to the tolerance of the resistor value. In all practical components, the marked value is a nominal value, and the real value will be the nominal value ± tolerance (1 to 20 percent, depending upon the quality of the resistor). Both of these problems can be overcome by making the feedback resistor variable. In most cases, the actual feedback resistor should be a series combination of a fixed resistor and a trimmer potentiometer (a multi-turn potentiometer). If the value of the potentiometer is approximately 10 to 20 percent of the overall value of R_f, and the fixed resistor has the rest of the resistance, a fine control over the circuit gain is possible.

In multistage amplifiers, it is usually sufficient to make the gain of only one stage variable. This arrangement allows us to trim the gain of the overall amplifier by trimming only one stage, which saves in trimpot cost. Note that an excessive number of trimpots will reduce the reliability of the circuit.

108

Chapter 11
Noninverting Followers

The inverting follower amplifier was first mentioned in Chapter 10. The transfer function of the inverting follower was developed from consideration of Ohm's law, Kirchoff's current law and the six basic properties of the ideal operational amplifier. In this chapter, we will use the same techniques to develop the concept of the noninverting follower.

Figure 11-1 shows the basic noninverting follower circuit. The feedback and input resistors are the same as in the inverting version, with the exception that the "input" end of the input resistor is grounded. The actual input signal is applied directly to the noninverting input of the operational amplifier. This arrangement offers a much higher input impedance than is possible in the inverting case. When the inverter is used, the input resistance is R_{in}. That sets the upper limit on the input impedance of the stage (again, recall the virtual ground). In the noninverting amplifier, however, the input impedance is set by the input resistance of the noninverting input. This is very, very high in real operational amplifiers and infinite in ideal op-amps.

CIRCUIT ANALYSIS

We want to find the transfer function of the noninverting follower circuit of Fig. 11-1. We may use the same approach as in the inverting case; namely, rely on Ohm's law, Kirchoff's law and the properties of the ideal operational amplifier.

The input potential, E_{in}, is applied directly to the noninverting terminal of the operational amplifier. By property No. 6 of the list given in Chapter 9, this means that we must treat the inverting input (point A) as if it were also at a potential of E_{in}. Input current I1, then, is found from:

$$I1 = E_{in}/R_{in}$$ **Equation 11-1**

Similarly, we are also able to find the value of the feedback current. But note that the voltage across the resistor is not simply E_{out}, but rather the difference between E_{out} and E_{in} (point A is at a potential of E_{in}):

$$I2 = \frac{E_{out} - E_{in}}{R_f}$$ **Equation 11-2**

From Kirchoff's current law, the two currents in this circuit sum to zero (there is no operational amplifier input bias current in our ideal case!), so we know that:

$$I1 = I2$$ **Equation 11-3**

Substituting Equations 11-1 and 11-2 into Equation 11-3 obtains:

$$\frac{E_{in}}{R_{in}} = \frac{E_{out} - E_{in}}{R_f}$$ **Equation 11-4**

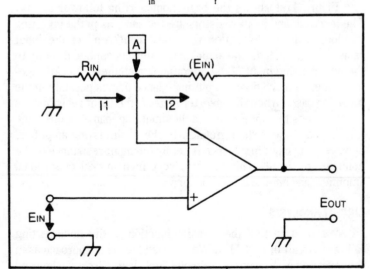

Fig. 11-1. Noninverting follower amplifier with gain.

Solving Equation 11-4 for the transfer function,

$$E_{out} = \frac{E_{in} R_f}{R_{in}} + E_{in} \qquad \textbf{Equation 11-5}$$

Rearranging, by factoring out E_{in};

$$E_{out} = E_{in} \left[\frac{R_f}{R_{in}} + 1 \right] \qquad \textbf{Equation 11-6}$$

The gain of the device is given by the expression, $A_v = (R_f/R_n) + 1$. At low gains, the "+1" term becomes very important, while at higher gains it is less important. Consider what happens at low gains. Assume that the feedback/input resistor ratio is 2:1. The gain of the circuit is:

$$A_v = \frac{2}{1} + 1 \qquad \textbf{Equation 11-7}$$

$$A_v = 2 + 1 = 3 \qquad \textbf{Equation 11-8}$$

But what if the ratio of the resistors is 200:1?

$$A_v = 200 + 1 = 201 \qquad \textbf{Equation 11-9}$$

In the low-gain case, the overall gain error, if the "1" factor is ignored, would amount to 33 percent, but in the high-gain case it is 0.5 percent. This is the reason why some texts erroneously list the gain of the noninverting follower as R_f/R_{in}. These books assume (dangerously) that the gain will be high—over 100.

UNITY-GAIN NONINVERTING FOLLOWERS

One specialized version of the noninverting follower is the circuit shown in Fig. 11-2. This circuit is called a unity gain follower because the voltage gain is 1. A unity gain noninverting voltage follower is made by connecting the output of the operational amplifier directly to the inverting input. In this circuit, the *beta* term (gain of the feedback network) is 1, so the gain with feedback will reduce to $A_{vol}/(1 + A_{vol})$. When A_{vol} is very high, as in the case of the operational amplifier, the apparent gain approaches unity.

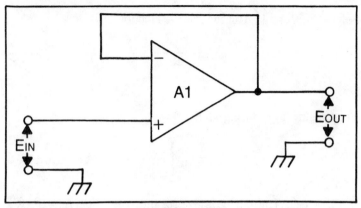

Fig. 11-2. Unity-gain noninverting follower amplifier.

Consider a practical example. An operational amplifier has an open-loop gain of 100,000 which is untypical. When the circuit of Fig. 11-2 is connected, the gain of the circuit is:

$$A_v = \frac{100,000}{1 + 100,000\,(1)} \qquad \textbf{Equation 11-10}$$

$$= \frac{100,000}{1 + 100,000} \qquad \textbf{Equation 11-11}$$

$$= 100,000/100,001 = 0.99999 \quad \textbf{Equation 11-12}$$

Few would argue that 0.99999 is close enough for 1.0 for practical work.

The main uses of the unity gain follower are *impedance transformation* and *stage buffering*. We know that the input impedance of the operational amplifier is typically over 500K ohms and may be as high as 1.5×10^{12} ohms. The output impedance, on the other hand, is typically less than 50 ohms. We may, therefore, use the unity gain follower to provide interfacing between a high-impedance source and a low-impedance load without loss of the voltage level.

The other typical application of the unity gain noninverting follower is in buffering two stages. A buffer amplifier is one that isolates two stages from each other. Some typical applications are between stages which need to see a constant—perhaps very high—load impedance. Actual loads that may tend to vary. Examples are oscillators feeding external circuits or additional stages of amplification and audio preamplifiers that feed power amplifier stages.

Chapter 12
Differential and Instrumentation Amplifiers

A differential amplifier produces an output that is proportional to the difference between two input signals. Both single-ended and push-pull output differential amplifiers are known. Both are available in integrated circuit form. Operational amplifiers also have the capability for differential amplification, and it will be toward op amps that most of this chapter is directed.

Figure 12-1 shows the basic discrete transistor differential amplifier. A stage like this will be found also in the input section of a IC operational amplifier device. Transistors Q1 and Q2 form the differential amplifier pair, while transistor Q3 forms a current source. When the bias network R3/R4 is connected to the V+ power supply, transistor Q3 will act as a constant current source. We will also find other uses for this transistor later in the Chapter.

We can make the differential amplifier work if we consider current I3 to be constant. We know from Kirchhoff's current law that:

$$I3 = I1 + I2 \qquad \textbf{Equation 12-1}$$

Because I3 is a constant, however, we can make certain statements regarding this relationship. We know that the equation must remain balanced, and I3 is constant. This means that a change in either I1 or I2 must be reflected as a change in the other current that reestablishes the equality with the constant I3. For example, if current I1 changes, there must also be a change in I2. The differential amplifier operation depends upon this phenomenon.

In the discussion of the differential amplifier shown in Fig. 12-1, we will make the assumption that emitter and collector currents in Q1 and Q2 are equal. This is not a tremendous departure from reality because, in fact, the emitter current will be 95 to 99 percent of the collector current; only a small base current causes the difference.

We must further establish that the output voltage is the difference between the collector potentials of Q1 and Q2: $V_o = V2 - V1$. Resistors R1 and R2 are equal, so collector currents I1 and I2 will also be equal under the condition where the input voltage E1 and E2 are equal. In that case, the collector voltages are equal, so output voltage V_o is equal to zero.

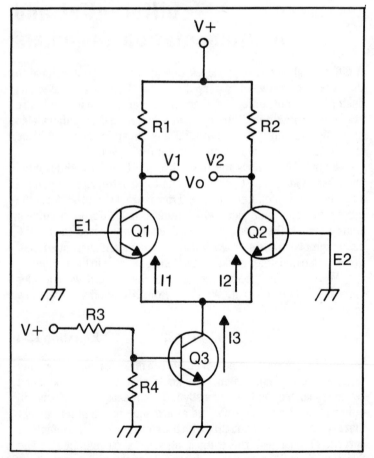

Fig. 12-1. Differential amplifier with current source.

But what happens when E1 $\neq$ E2? Let us say that E1 is greater than E2. In that case, the collector current of Q1 *rises* (I1 goes up), so the collector current of Q1 (I2) must go down. What happens to the voltage at the output? Increasing the collector current of Q1 causes a greater portion of the collector supply voltage, V_{cc}, to be dropped across R1, causing V1 to go down. At the same time, the reduced value of I2 causes the voltage drop across R2 to decrease, forcing V2 up. Because there is an inequality of the collector voltages, differential output voltage V_o is no longer zero.

The maximum voltage swing allowed V_o is the full potential V_{cc}, instead of just $\frac{1}{2}V_{cc}$, as might be expected in a class-A single-ended amplifier. The full voltage swing is obtained because of the differential nature of the output voltage. It will occur at a maximum when one of the transistor collectors is at V_{cc}, and the other is at $V_{ce(sat)}$ (close to zero volts).

The circuit of Fig. 12-1 can be used in a wide variety of applications where differential input signals must be processed. It is also sometimes seen in rf amplifiers and in mixer stages for superheterodyne receivers, balanced modulators for single-sideband generation and in other frequency translation applications. We can use the circuit as a frequency converter by applying the local oscillator signal to the base of Q3 and the rf signal across the differential pair.

The differential amplifier is used as an rf amplifier by connecting an appropriate input tank across the bases of Q1 and Q2. For manual gain control operation, we will bias Q3 to some fixed point for maximum gain and then reduce the gain by reducing the bias when needed. We would connect the open end of R3 to a variable voltage source, such as a potentiometer, rather than V+. Note that this same point also makes a good automatic gain control (agc) point. We would develop a DC signal that is proportional to the strength of the i-f signal and then apply it to the base of Q3. This voltage would counteract the normal bias voltage.

For use as a balanced modulator for SSB generation, we would apply the carrier oscillator signal to the base of Q3 and then apply the audio speech signal in push-pull across the bases of Q1 and Q2. The output would be a tank circuit across the collectors of Q1 and Q2, tuned to the frequency of the oscillator. The collector currents of Q1 and Q2 would cancel each other in the tank circuit, except when the push-pull audio drive signal unbalanced the signal.

SINGLE-ENDED OUTPUT

The standard differential amplifier of the previous example

was equipped with a push-pull output. We frequently need a single-ended output from a stage that has differential inputs. Figure 12-1 shows such a stage. It is basically the same as the stage of Fig. 12-1, except that just one collector (Q2) is being used as the output. In this circuit, it is mandatory that the emitter current source be constant: I3 must be a constant current source. If a circuit such as Fig. 12-1 is used, then, the base of Q3 should be connected to a fixed bias source.

There are two inputs in the circuit of Fig 12-1. The base of transistor Q1 is called the *noninverting input*, and it is denoted with the + symbol. The base of transistor Q2 is called the *inverting input* and is denoted by the − symbol. These two inputs have equal but opposite effect on the output, V_o. We discussed this circuit in Chapter 9 on basic operational amplifiers, but we should review its operation here for emphasis. Recall that I3 is constant, and I3 = I1 + I2. If we cause a change in I1, there will also be a change in I2. Similarly, when there is a change in I2, there will also be a change in I1.

The output voltage, V_o, is the collector voltage of transistor Q2. It will always be the V+ potential, less the voltage drop across resistor R2. If current I2 increases, the voltage drop across R2 also increases, leaving less of the total collector voltage from V_o. Similarly, when the collector current I2 decreases, the voltage drop across R2 also decreases, forcing the output voltage V_o upward.

Inverting Operation

For the moment, assume that the base of Q1 is either zero or fixed at some quiescent bias (nearer the truth). We will apply a sine wave voltage to the base of Q2. What happens? On the positive half of the sine wave, the collector current of Q2 *increases*, so V_o will *decrease*. Similarly, when the sine wave is on the negative half of its cycle, the collector current of Q2 decreases, forcing V_o to increase. Note the inverted action at the output terminal: Positive-going input voltages cause a negative-going output voltage, and vice versa. Hence, the base of Q2 is called an *inverting input*.

Noninverting Operation

For noninverting operation, we will hold the inverting input at a constant potential and apply the sine wave to the noninverting input, the base of Q1. On the positive half of the sine wave, current I1 will increase. To maintain the constancy of I3, therefore, I2 will decrease. This will force V_o to increase. Similarly, on the negative half of the sine wave, collector current I1 will decrease, forcing I2 to increase. This will increase the voltage drop across R2, thereby reducing V_o. In this mode of operation, a positive-going signal at the noninverting input causes a positive-going output signal, and vice versa. Hence, the base of Q1 can indeed be called the *noninverting input*.

OPERATIONAL AMPLIFIER DIFFERENTIAL AMPLIFIERS

Most operational amplifiers have two inputs: inverting and noninverting. These inputs both see the same voltage gain, but they have opposite effects at the output terminal. If the same voltage is applied to both inputs, then, the output voltage contributions from each input cancel each other: The output is zero. We call a voltage that is applied to both inputs a *common-mode voltage*.

The output potential in a differential amplifier is proportional to the difference between two input potentials. If we label the output voltage, V_o, and the two input potentials, V_1 and V_2, the transfer equation for the differential amplifier is:

$$V_o = A_{vd}(V2 - V1) \qquad \textbf{Equation 12-2}$$

where V_o is the output potential, V_1 is the potential applied to the inverting input, V_2 is the potential applied to the noninverting input, and A_{vd} is the differential voltage gain of the circuit.

We can get a glimpse of one of the main uses of the differential amplifier by considering Equation 12-2 when $V_1 = V_2$. The differential input voltage becomes zero (V2 − V1 = 0), so the output voltage also becomes zero. This is known as *common-mode rejection*. It is a very handy phenomenon when a large interfering signal affects both inputs equally. Such is often the case in certain instrumentation applications.

The electrocardiograph amplifier is designed to pick-up weak signals from the surface of the patient's body. These signals have peak amplitudes of 1mV or less. Yet the interfering 60-Hz signal from nearby power lines will have amplitudes of up to several volts. The common mode rejection of a good differential amplifier is enough to suppress this signal and recover the ECG signal that is several thousand times weaker.

A simple DC differential amplifier using just one operational amplifier is shown in Fig. 12-3. This diagram includes the various voltages at the input to better appreciate their relationship.

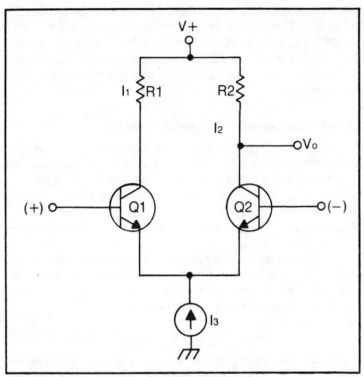

Fig. 12-2. Equivalent circuit for differential amplifier.

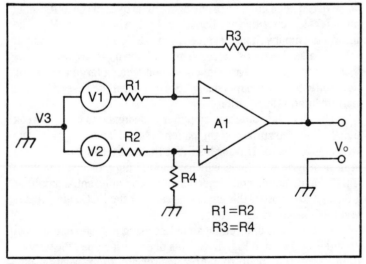

Fig. 12-3. Simple DC differential amplifier based on op amp.

Voltage V3 is the common-mode voltage, while V1 and V2 are the signal voltages applied to the inverting and noninverting inputs, respectively. The differential input voltage will be V2– V1.

The voltage gain of this circuit is given by the following expression, provided that R1 = R2, and R3 = R4:

$$A_{vd} = R3/R1 \qquad \text{Equation 12-3}$$

Note that Equation 12-3 is essentially the gain of an inverting follower stage. We must be certain to ensure that the resistor equalities just mentioned are maintained. There will be a substantial gain error, and the common-mode rejection will be lost if these equalities are not maintained properly. It is sometimes the practice to make resistor R4 either a potentiometer or a series combination of potentiometer and a fixed resistor. This will allow us to trim out the normal differences in resistor value and optimize the common-mode rejection of the amplifier. Such a control is sometimes marked *common-mode adjust* (CMA) or *common-mode rejection*.

COMMON-MODE REJECTION RATIO

If the differential operational amplifier were perfectly ideal, the equal and opposite action of the two inputs would completely suppress all common-mode signals. But practical operational amplifiers are nowhere near ideal: They have lots of errors, and one of them is an imperfect common-mode rejection ratio. There will be some output due to common-mode input voltages, so there is necessarily a common mode gain, A_{vcm}, which is defined as:

$$A_{vcm} = \frac{V_{ocm}}{V_{icm}} \qquad \text{Equation 12-4}$$

where A_{vcm} is the common-mode voltage gain, V_{ocm} is the output voltage due to a common-mode input signal, and V_{icm} is the common-mode input signal voltage (V_3 in Fig. 12-3).

The common-mode rejection ratio (CMRR) is a measure of the ability of an amplifier to reject common-mode signals. CMRR is defined as the voltage gain to the common-mode gain:

$$CMRR = \frac{A_v}{A_{vcm}} \qquad \text{Equation 12-5}$$

where *CMRR* is the common-mode rejection ratio, A_v is the voltage gain, and A_{vcm} is the common-mode voltage gain (Equation 12-4).

The CMRR is often seen expressed in decibel form, which is defined as:

$$CMRR_{dB} = 20 \log_{10}(A_v/A_{vcm})$$ **Equation 12-6**

Most operational amplifiers have the CMRR expressed in decibels. Table 12-1 shows some of the conversions for CMRR expressed in dB.

The maintenance of a high CMRR is necessary to reduce interference from external electrical fields. The 60-Hz power mains field is probably the single biggest pain in the neck seen in this area. Several factors will deteriorate the CMRR, and these must be addressed in differential amplifier circuits:

☐ Poor component matching (R1 ≠ R2 and R3 ≠ R4 in Fig. 12-3) that results in a poor initial CMRR for the amplifier. This is solved partially by using a matched resistor and a good quality operational amplifier. Optimization of CMRR also requires a potentiometer for R4 (Fig. 12-3).

☐ Mechanical anomalies in the connections.

☐ Improper or inappropriate connection schemes.

☐ Generation of differential signals from common-mode signals. This issue is addressed in Chapter 25.

IMPROVED DIFFERENTIAL AMPLIFIER

The amplifier circuit shown in Fig. 12-3 offered only one appealing attribute: it is cheap! It can be built from only a single IC operational amplifier, but it also suffers from some defects. One is a limitation on input impedance at high gains. The gain of the circuit

Table 12-1. CMRR Conversion.

CMRR	CMRR in dB
1,000,000:1	120 dB
100,000:1	100 dB
10,000:1	80 dB
1,000:1	60 dB
100:1	40 dB

is set by the ratio of the feedback to the input resistors, as is normally the case in operational amplifier circuits. But this is the very thing that will limit the input impedance and—necessarily—the gain. A certain minimum input impedance must be maintained all of the time, lest the preceding stage be loaded. This will impose a minimum value on R1/R2. Similarly, there is a maximum value for R3/R4 because of problems with noise, input bias currents and the ease of obtaining high-value (multimegohm) resistances. Because of these limitations, there is a maximum gain that is attainable in any given model op amp using the circuit of Fig. 12-3. An improvement is offered by the *instrumentation amplifier* circuit of Fig. 12-4.

Amplifiers A1 and A2 form the input stage of the instrumentation amplifiers. These devices should be premium models and work best if the input impedance is very high (the biMOS and biFET types). The input amplifiers are connected in the noninverting configuration.

The output stage (A3) is connected in the DC differential amplifier configuration of Fig. 12-3. In the simplest case, where the four resistors of this stage are equal (R4 = R5 = R6 = R7), the stage gain is unity. The overall gain of the instrumentation amplifier will therefore be:

$$A_{vd} = \frac{2\,R3}{R1} + 1 \qquad \textbf{Equation 12-7}$$

We assume in Equation 12-7 that R2 = R3 and that R3 = R4 = R5 = R6. It is interesting to note that the common-mode rejection ratio of the instrumentation amplifier is not materially affected by matching errors between R2 and R3; only the gain is affected. However, a mismatch of R2 and R3 will result in a gain error.

The situation created when the gain of the output stage is unity is wasteful, to say the least. If the gain of A3 is greater than unity, Equation 12-7 must be rewritten in the form:

$$A_{vd} = \left[\frac{2R3}{R1} + 1 \right] \left[\frac{R7}{R5} \right] \qquad \textbf{Equation 12-8}$$

☐ **Example:**
Calculate the differential voltage gain, A_{vd}, of an instrumenta-

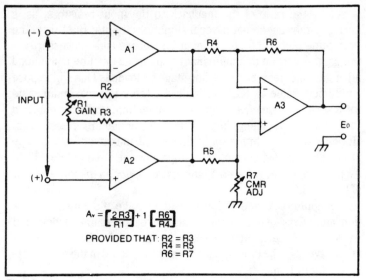

Fig. 12-4. Instrumentation amplifier.

tion amplifier when the following resistor values are used: R3 = 47K ohm, R1 = 1K ohm, R5 = 3.9K ohm, and R7 = 22K ohms.

$$A_{vd} = \left[\frac{(2)(47K\ ohm)}{(1K\ ohm)} + 1\right]\left[\frac{(22K\ ohms)}{(3.9K\ ohms)}\right]$$

$$= (94 + 1)(5.6)$$

$$= (95)(5.6) = 532$$

It is interesting to note that a general expression for the gain of the basic instrumentation amplifier circuit can be written:

$$A_{vd} = \frac{R7(R1 + R2 + R3)}{R1R6} \qquad \textbf{Equation 12-9}$$

Which will remain valid, curiously enough, not just under certain sets of resistor equality values, but when the *ratios* of resistor pairs are equal: R7/R6 = R5/R4.

The circuit for a second instrumentation amplifier, using but two operational amplifiers, is shown in Fig. 12-5. This circuit is a little less complex than the previous circuits but suffers from being a little more sensitive to component errors. As in the previous circuit, input signal V_i is applied across the noninverting inputs of A1 and A2. The gain is set by R3. The expression for the gain of this stage is:

122

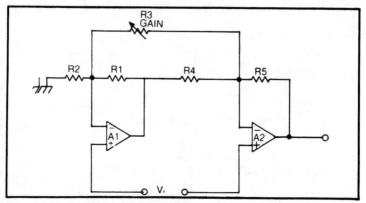

Fig. 12-5. Alternate instrumentation amplifier.

$$A_{vd} = \frac{R_2(2R_1 + R_3)}{R_1 R_3} + 1 \qquad \textbf{Equation 12-10}$$

Provided that R1 = R4, and R2 = R5.

DIFFERENTIAL AMPLIFIER APPLICATIONS

The list of potential applications for a differential amplifier could conceivably be very long, indeed. We have already touched on the biomedical applications. These are not unique, incidentally, but merely representative of a larger class of applications where we have a large interfering signal in the presence of a small desired potential. We must conspire to make the desired signal differential and then allow the interfering signal to be common mode. The result, as in the case of the EEG and ECG amplifiers, will be to reduce the level of the noise signal considerably.

Another principal application for differential amplifiers is processing the signal from a transducer. Many transducers are either Wheatstone bridge circuits in their own right, or can be

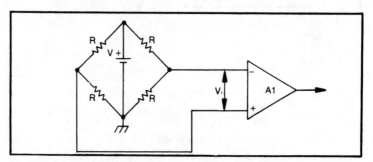

Fig. 12-6. Differential amplifier used as to amplify output of Wheatstone bridge.

123

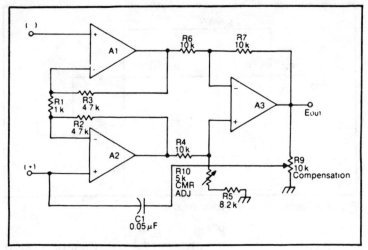

Fig. 12-7. Practical instrumentation amplifier.

connected into a Wheatstone bridge with other circuit elements.
Figure 12-6 shows a differential amplifier handling the output of the
Wheatstone bridge transducer. A typical Wheatstone bridge will
consist of four arms, with a resistance, R (we are considering only
the case where all resistors are equal). The bridge is excited with a
precision voltage, V. When the transducer is not under any strain,

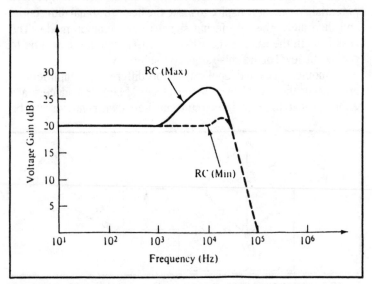

Fig. 12-8. Frequency response of amplifier.

124

all four resistors have equal value, so the voltages applied to the two inputs of the differential amplifier are equal. But when the bridge is strained, the values of the resistors change in a manner that will cause the input voltages to become nonequal, making V_i nonzero. An output occurs, therefore, only when the transducer bridge is strained.

PRACTICAL DIFFERENTIAL
AMPLIFIER FOR BIOPHYSICAL MEASUREMENTS

Investigators in the life sciences often require differential amplifiers with gains anywhere from times-10 to times-100. They will frequently be required to connect a high-impedance source (animal or human biopotential) to an amplifier through some coaxial cable to reduce the amount of noise artifact. The capacitance of the coaxial cable, coupled with the high impedance involved, can roll off the frequency response of the amplifier considerably. Consider a 5′ piece of coaxial cable. It will normally have a capacitance of about 30 pF/ft, so the section would have a capacitance of 150 pF. When the input/source impedances are in the range of 1 megohm, the frequency response of the amplifier falls off to 1 to 3 kHz. This is not significant in some studies but becomes important when looking for high repetition rate pulses that each have a fast rise time.

Figure 12-7 shows the circuit for an instrumentation amplifier that is designed to compensate for the capacitance of the input cable. The basic instrumentation amplifier portion of the circuit consists of three operational amplifier devices. If we are not terribly concerned with input impedance, we may use 741-family devices for all three op amps. If we need the high impedance preserved (the usual case when measuring biopotentials), though, we must use a premium-grade device with a high-impedance front end. This requirement is best met with biMOS (CA3140/CA3160 devices), biFET (LF156, etc.) or superbeta (Darlington amplifier) input stages.

The compensation for the circuit capacitance is provided by potentiometer R9 and feedback capacitor C1. The result of varying this potentiometer is shown in Fig. 12-8. The response shows a definite peaking when the RC time constant is maximum and will compensate for the rolloff at the input stage of the instrumentation amplifier. Note that higher gains will necessitate changing the values of RC network, or oscillation will result at some setting of the potentiometer.

Chapter 13
Isolation Amplifiers

The isolation amplifier is a relatively new device in the armamentarium of the electronic circuit designers, but its range of applications are expanding. In those places where the use of an instrumentation amplifier is indicated, absolutely no substitute is satisfactory. In certain industrial instrumentation applications and in many biomedical applications, the isolation amplifier is needed.

The principal use of the isolation amplifier is where the amplifier input must be *isolated* (hence the name) from the amplifier output. In some industrial applications, this is required where the amplifier will be in some electrically sensitive region and the transmission of an instrumentation analog data signal to a remote location is desired.

But the major use of the isolation amplifier is in biomedical electronics. It has been learned over the past several decades that patients in hospitals are more susceptible to electrical injury than other people. The current required to kill or seriously injure a person with intact skin is (depending upon whom you accept as the authority) 50 to 500 mA (100 to 200 mA seems to be a most satisfactory common ground). But to the patient with a breech in the protective layer of skin, the susceptible current drops down to some value in the under 500 *microampere* range. Various authorities claim that the threshold current permissible for a patient with a skin wound, IV needle, or in-dwelling catheter is any value between 10 uA and 150 uA (again, authorities disagree). But leakage currents from the AC power mains, which are of no importance on normal amplifiers, can easily exceed these levels, especially if the ground wire from the chassis of the amplifier to the power mains ground is interrupted. The leakage currents are caused by induction and capacitive coupling between the AC power mains wires coming into a chassis, and the chassis itself. Normally, these currents are drained off to ground and become harmless. But when the ground wire is broken, or when the tension on the plug ground blade by the wall socket is too low, a high resistance to ground will exist and a leakage current will flow. There is much discussion on how much leakage current is acceptable in medical equipment, but the general consensus seems to be "the less, the better." In most cases, however, the lower limit of 10 uA is a little difficult to meet in amplifiers without isolation techniques.

There are two sources of undesired current: one is the power line leakage mentioned above, while the other is currents passed down the lead wires from the patient's body to the amplifier input. The value of a high-impedance input is readily apparent (the patient resistance is 10K ohms or more!), but the need for isolation from the AC power mains is apparent mostly to the biomedical engineer. Commonly available isolation amplifiers, using any of several different technologies, can boast input-to-AC-mains impedances of 10^{12} ohms. That's 1 terraohm, or 1,000,000,000,000 ohms! In this chapter, we will discuss some of the more commonly available isolation amplifiers. If you have an interest in pursuing biomedical electronics, refer to TAB book No. 930, *Servicing Medical & Bioelectronic Equipment.*

ISOLATION AMPLIFIER SYMBOL

The standard amplifier symbol used extensively in electronics is a triangle on its side. The output is usually specified as the apex of the triangle facing horizontally, while the inputs are along the vertical side opposite the output. This has become the standard symbol for operational amplifiers, other integrated circuit amplifiers and amplifiers in general (when viewed as a black box). This symbol is modified in Fig. 13-1 to make an unofficial isolation amplifier symbol. Although there are several versions of the isolation amplifier symbol are known, the two shown in Fig. 13-2 are the most common. The symbol in Fig. 13-1A is probably the oldest and consists of the triangle operational amplifier symbol broken into two parts. More recently, we have seen the symbol of Fig. 13-1B. This symbol is a little more appealing to many and consists of the broken triangle with dotted lines to indicate the isolation between input and output sections. A wiggly arrow indicates that—despite the isolation—signal flows.

Figure 13-2 shows an isolated electrocardiograph (ECG) amplifier used in a popular medical cardiac monitor. The isolated amplifier portion of the circuit is shown in the upper box. The ECG signal is picked up from the surface of the patient's body through connector BAED and applied to the input of a differential instrumentation amplifier. A right leg drive amplifier is the same as the common-mode, or guard, amplifier used in certain nonmedical instrumentation amplifiers. The function of the right leg amplifier, like that of the common-mode amplifier, is to supply a counter-EMF to the common signal point that is equal to, but opposite in phase with, the common-mode signal from the 60-Hz power mains. This signal can induce up to several volts of signal into the leads, while the ECG signal is only 1 mV on peaks, with many salient features are 100 μV. The gain of the amplifier may be moderate (times 100 to times 500) if there is amplification following the isolated stage, or high (times 1000 to times 2000) if there is no significant amplification following the isolated stage.

The *gain cal* adjustment is used to provide a 1 mV DC level from a front panel switch. It is normal procedure in ECG work to

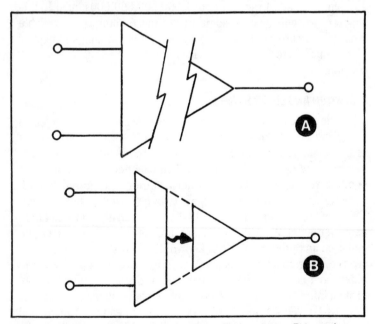

Fig. 13-1. Symbols used for isolation amplifiers. Either one (A or B) is used.

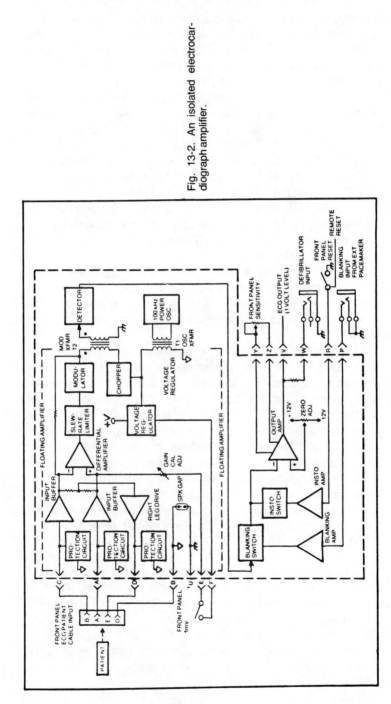

Fig. 13-2. An isolated electrocardiograph amplifier.

129

adjust the oscilloscope of paper strip-chart recorder to produce two divisions (each division either 5 or 10 millimeters) deflection for a 1 mV signal. The amplified ECG signal is passed through a slew-rate limiter, which is a filter circuit that has the effect of reducing the maximum rate of change of the output signal.

The key to isolation in this design is the 100-kHz power oscillator. The output of this oscillator is passed to the isolated section through a 100-kHz transformer. The design of this transformer is controlled so that it works efficiently at 100 kHz but poorly at 60 Hz (the power mains frequency). This means that little, if any, of the 60-Hz signal from the power mains enters the amplifier section that interfaces with the patient.

Power to the isolated amplifier section is obtained from the 100-kHz signal. This signal is rectified and filtered, then fed to a voltage regulator to provide the V+ DC potential required by the amplifiers and other circuitry.

The ECG signal is transmitted to the nonisolated section through a modulation transformer that also operates at 100 kHz. The ECG signal is applied to a chopper modulator stage on the isolated side, where it will amplitude modulate the 100-kHz carrier signal.

In this particular circuit, the modulated signal containing the ECG information is transferred across the modulation transformer to a simple envelope detector. This is the simplest form of detector that is sensitive to amplitude modulation of a carrier. If an FM carrier is used, an FM phase detector is required for the detector stage. In many cases, however, you will see a synchronous phase detection scheme used to demodulate the amplitude modulated data. This method has the advantage of cancelling the even-order distortion products in the demodulated waveform. In the case where a phase-sensitive synchronous detector is used, a sample of the 100-kHz oscillator signal from the nonisolated side of the oscillator transformer is fed to the detector to serve as the reference.

The circuit of a typical phase sensitive detector is shown in Fig. 13-3. This circuit shows two pnp bipolar transistors as a pair of push-pull electronic *switches*. In more modern circuits, including several that are used extensively in isolation amplifiers, CMOS switches are often used instead of the bipolar transistors shown. In that case, a device such as the CA4066 CMOS switch would be used, and the reference signal applied to the control inputs of the switch would have to be a square wave. Other models use varactor diode bridges for the same purpose.

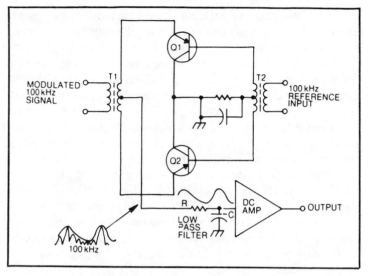

Fig. 13.3. Transistor synchronous detector.

The operation of this circuit is relatively straightforward. On the positive half of the reference signal, one transistor will be turned on and the other is off. This will cause one end of the input transformer (T1) to be grounded, permitting its polarity with respect to the center tap of the secondary to govern the polarity of the output signal. On the other half-cycle of the reference signal, the other transistor will be on, causing the opposite end of the transformer secondary to be grounded. The effect is to keep the polarity of the output signal always positive or negative, depending upon the phase of the reference signal. This is, then, equivalent in all respects, to the process called *full-wave rectification*. Only the amplitude of the output signal varies with the input signal applied to the primary of T1. In the CMOS switch version, incidentally, the reference signal drives the switch control inputs, while the analog channels of the switches connect each end of the transformer to the load (Fig. 13-4). In the CMOS version, the channel resistance is low when the switch is on and very high when the switch is on. This effectively creates the same action as noted above.

The rest of the circuitry on the nonisolated side of the ECG amplifier shown in Fig. 13-2 is specific to ECG amplifiers and may or may not be found in other types of amplifier. The blanking amplifier is merely an electronic switch that will turn off the amplifier during pacemaker spikes. The pacer is an electronic pulse generator that creates a sharp pulse artifact in the ECG

131

signal, and this stage is used to remove it from the displayed signal. The INSTO is a special circuit that allows the amplifier to recover from extreme overloads to the input. Patient movement produces bioelectric potentials from the skeletal muscles that are several times the amplitude of the ECG signal. This will tend to make the amplifier saturate. If a defibrillator (an electrical shock device used to correct a particularly deadly form of cardiac arrhythmia) is used on the patient, the input will see a high-voltage pulse that may run almost to a kilovolt—1,000,000 times the normal input signal amplitude! The INSTO will help the amplifier recover in time for a physician or other qualified rescuer to see if the jolt did its job...or whether the victim is gonna get it again. The protection circuits on the isolated side of the circuit are also used for defibrillators, but they are needed to protect the delicate input circuitry from being sent into orbit by the defibrillator blast. Such protection is usually in the form of series, current-limiting resistors and neon glow lamps (NE-2) across the input lines.

The output of the detector in both forms of circuits (synchronous or nonsynchronous) will be a series of rectified pulses whose amplitude variations carry the desired analog waveform data. We must get rid of the 100-kHz component before the data can be displayed. This is done in a low-pass filter that has a cutoff frequency above the data frequency (100 Hz in ECG amplifiers and rarely more than 1000 Hz in instrumentation amplifiers), yet is considerably lower than the 100-kHz carrier frequency. The result is to chop off the residual 100-kHz component, leaving only the analog waveform applied to the input of the isolated amplifier. A

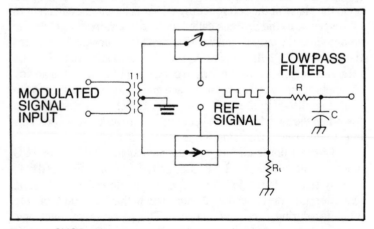

Fig. 13-4. CMOS switch version of a synchronous detector.

DC amplifier at the output of this circuit is used in most cases to build up the amplitude of the analog waveform and to provide any level-shifting needed to be rid of any DC offsets picked up in the process.

The circuit discussed so far is merely one of the most common forms of isolation amplifier; it is not totally universal, however. There are at least two other types: optically coupled and current-starvation.

OPTICALLY COUPLED ISOLATION AMPLIFIERS

The transformer coupled isolated amplifier is one of the earliest types known, probably because design engineers in the medical electronics field had plenty of experience with transformer coupled techniques in the design of AC carrier amplifiers for pressure measurements. But in the 1960s a neat little device called a *light-emitting diode* (LED) was introduced. This semiconductor device would generate a light when a DC potential was applied across its pn junction. In most cases, the LED output light was red. If the voltage applied to the LED were made to vary in amplitude, then the output light would also vary, although linearity is sometimes suspect. Alternatively, if the frequency of the applied voltage were varied (sometimes with a DC offset), the light emitted from the LED would contain these frequency variations. Hence, the LED can be used to transmit either amplitude-modulated or frequency-modulated variations.

An optoisolator is an integrated circuit device containing an LED placed opposite either a phototransistor or photoresistor; the transistor is more common. We can, therefore, transmit light variations (AM or FM) across the open space between the LED and the phototransistor. Optoisolators in 6-pin miniDIP IC cases can have 1-terraohm resistance between the LED and the transistor (the series leakage resistance of the plastic or ceramic forming the IC package).

Figure 13-5 shows an isolation amplifier based on the LED optoisolator just described. The amplifier is a typical differential amplifier and will be an operational amplifier in all but a few cases. DC power to the operational amplifier is supplied through a DC-to-DC converter. This is essentially the same as in the previous case, except that the converter can be supplied as a separate module, a hybrid function module. It consists of a 50-kHz to 300-kHz power oscillator and a transformer. It operates from the V− and V+ power supplies on the nonisolated side of the amplifier and produces isolated DC (at low current) on the isolated side of the dotted line. The isolated DC power is designated $V^{*}-$ and $V^{*}+$ in the schematic of Fig. 13-5. The analog input signal is applied to amplifier A1, which builds its amplitude up to a level that will satisfactorily drive the modulator, a *voltage-controlled oscillator* (VCO) stage.

A VCO is an oscillator that produces an output frequency that is a function of an applied control voltage. If the control voltage is an analog signal, the oscillator output will be frequency modulated about its $V_{in} = 0$ frequency. The FM output of the VCO is applied to the light-emitting diode and is then transferred across the isolation barrier to the rest of the circuitry. The detector will have to be sensitive to phase variations (FM is an angular modulation). In most cases, because the FM carrier is audio, the *phase-locked loop* (PLL) or digital pulse-counting FM detector techniques are used for demodulation.

The best signal-to-noise ratio is obtained in audio FM when the carrier frequency and deviation are related such that the deviation is 75 percent of the carrier frequency. For example, if a 1000-Hz carrier (f = 1 kHz when V=0) frequency is used, the deviation will be ±750 kHz. The carrier will swing from 250 Hz to 1750 Hz on amplitude peaks. This specification also makes it relatively easy to follow with a untuned detector, such as the pulse counting circuit.

Note the grounds in this and the previous isolation amplifier. We want the circuit to be totally isolated where possible from the AC power mains. The mains are *ground-referenced.* The grounds on

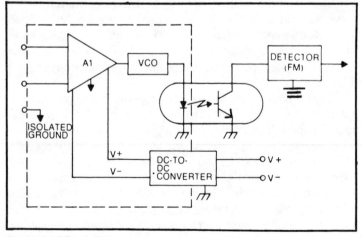

Fig. 13-5. LED isolation.

the nonisolated side of the dotted line are connected to the power line earth ground so are shown as the normal three-bar ground symbol. The grounds on the other side are actually not grounds at all, but a floating common connection, or *counterpoise ground*. These grounds are usually shown as an open triangle to differentiate with the earth ground.

The rest of the circuit in an optically coupled isolation amplifier are pretty much the same as in the previous circuit. If the device is an ECG amplifier, the same types of stages will be needed. If the amplifier is for general instrumentation purposes, gain control, frequency response tailoring and the level shifting functions will probably be found in the post-isolation side of the amplifier.

CURRENT-STARVATION ISOLATED AMPLIFIERS

The basic circuit for a clever isolation amplifier is shown in Fig. 13-6. On first inspection, it does not appear that the signal from the output side of the amplifier has a way across the isolation barrier (dotted line). Let's look a little closer though. Transistors Q1 and Q2 form a 250-kHz power oscillator to provide isolated power to the amplifier. So far, there is nothing different from the other two designs. But this oscillator can only supply a light load equal to the quiescent current drain of the isolated amplifier. When an input signal causes the output of the amplifier (A1) to vary, the current loading of the transformer secondary varies. This is reflected to the primary side of the circuit. These variations are seen as current changes in the emitter resistors of the power oscillator, which in turn look like voltage changes across resistor R1. The output signal, then, is a varying voltage that is proportional to the loading of the transformer secondary via the V^*+ power line. The output signal across resistor R1 will have a constant DC offset due to the emitter currents of the oscillator transistors, so some form of level shifting is needed in the DC amplifier that normally follows the output line.

ISOLATION AMPLIFIER APPLICATIONS

The isolation amplifier is used mostly in biomedical applications and may indeed have been designed primarily out of the patient safety needs of the biomedical electronics industry. The range of possible instrumentation applications is almost unlimited, however, and might not necessarily involve only rugged environment applications. Basically, the isolation amplifier may be used anywhere that a high degree of isolation (10^{12}) from the AC power mains is required. Applications listed in manufacturers' catalogs include three-wire instrumentation transducer circuits, isolated thermocouple and TRD instrumentation, chemical and industrial process control, current loop and current shunt applications and ground loop elimination (perhaps the most important application in nonrugged environments). Of course, there are also the biomedical applications.

A COMMERCIAL PRODUCT

Figure 13-7 shows the block diagram of a commercial hybrid isolation amplifier: the Burr-Brown 3456. The hybrid form of construction uses both IC chips and a highly specialized form of substrate wiring to form highly efficient and complex electronic circuit modules with properties that are difficult to achieve with either impedance and high differential gain in one package. You would recall that all this is almost impossible in a single op amp differential circuit. The gain of the differential instrumentation amplifier is given by the equation $(2R_f/R_{in}) + 1 = A_v$. In this circuit, the gain can be set by a single external resistor because the feedback resistances are set at 12.5K ohms by the manufacturer. This leaves only the input resistor, which reduces the gain formula to:

$$A_v = G = \frac{25K \text{ ohms}}{R} + 1 \qquad \textbf{Equation 13-1}$$

☐ **Example:**

Find the gain of an amplifier such as the one shown in Fig. 13-7 if the external resistor has a value of 200 ohms.

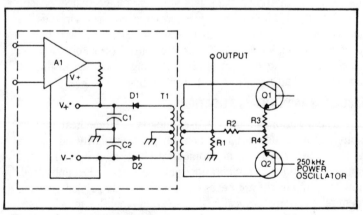

Fig. 13-6. Current starvation isolation.

136

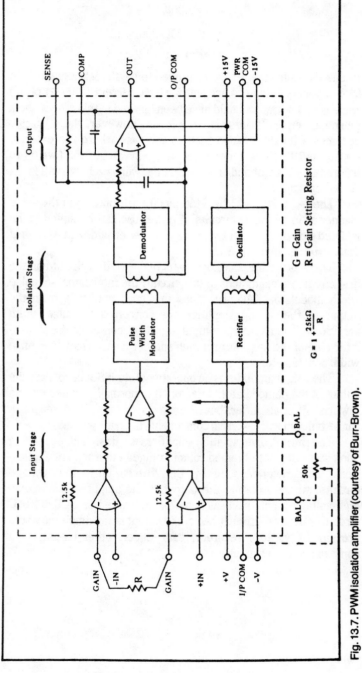

Fig. 13.7. PWM isolation amplifier (courtesy of Burr-Brown).

137

$$A_v = (25{,}000/R) + 1$$
$$= (25{,}000/200) + 1$$
$$= 125 + 1 = 126$$

In this example, the +1 factor proved to be almost negligible; in fact, it was only able to account for 0.81 percent (0.0081) of the total gain. This factor could have been ignored in this case without much gain error. If the gain were lower, however, the +1 factor becomes a lot more important. In the case where the external resistor is 10K ohms, for example, the gain will be 3.5. If we try to ignore the +1 factor and pretend that the gain is 2.5, we will find a 29 percent gain error.

The DC power for the 3456 module is obtained in the same manner as in the other circuits. A power oscillator coupled to the isolated side of the circuit via a transformer provides an AC signal that can be rectified and filtered.

There is a difference, however, between this circuit and the first circuit we examined. In this circuit, the modulator is a pulse width modulator. The variations in signal amplitude from the output of the analog amplifier are converted to pulse width variations. The modulator output signal is coupled to the nonisolated side of the amplifier through another transformer to a pulse width demodulator.

The output amplifier is an operational amplifier coupled to the output of the demodulator/detector. It is normally connected as a low-pass filter amplifier, but the gain and the frequency response can be tailored with external components to your own needs.

The Burr-Brown isolation amplifiers are designed, depending upon style, for both biomedical and rugged environment applications. One recommended by Burr-Brown is in high-voltage circuits, where isolating the output of the amplifier from some high-voltage input, such as a motor controller or SCR/TRIAC circuit, is desired. In such cases, the output of the amplifier would presumedly go to a recorder, data logger, other circuit, or a communications channel.

Chapter 14
Operational
Amplifier Problems

The ideal operational amplifier of Chapter 9 does not really exist. Some premium devices on the market will approach the ideal to a startling accuracy, but they *all* fall short of the performance of the true ideal operational amplifier. Fortunately, many of the problems are easily recognized and may be dealt with using standard circuit techniques.

One principal departure from the ideal in real operational amplifiers is in the area of the input bias current, I_o. This current will be zero in the ideal case because of ideal property No. 2: $Z_{in} =$ infinity. The ideal operational amplifier input will neither sink nor source any current. But the real operational amplifier will have some small input current due to the requirement to bias the input transistors. This current will create an offset output voltage that is equal to $-I_o \times R_f$. In low-gain circuits, this offset voltage is not terribly large and can often be ignored. But as the circuit gain increases, then so does the magnitude of the output offset voltage that is due to these bias currents. In the circuit shown in Chapter 10, which has a 500K ohm feedback resistor, the output voltage caused by a small 10-μA bias current would be on the order of (0.000010 A) (500,000 ohms), or 5 volts. This voltage would be seen as a DC potential about which the correct output potential varies. A 5VDC component is unacceptable, so there must be some means for getting rid of the effects of the input bias current.

One method is shown in Fig. 14-1: Use a *compensation resistor* in the noninverting input. The bias currents in the two inputs are approximately equal. Further, that the current in the inverting input produces a voltage at the inverting input that is equal to the product of the input bias current, I_o, and a resistance equal to the parallel combination of R_f and R_{in}. We can counteract the effect of this voltage by placing an equal resistance between the noninverting input and ground, as in Fig. 14-1. The value of this resistance, R_c, should be approximately the parallel combination of R_{in} and R_f:

$$R_c = \frac{R_{in}\,R_f}{R_{in} + R_f} \qquad \textbf{Equation 14-1}$$

We can usually tolerate a little slop in R_c, so we obtain a resistor that is the nearest standard value to the calculated value. For example, if the feedback resistor is 100K ohms, and the input resistor is 1K ohm, the correct value for R_c would be 99K ohms. Little is sacrificed, however, if we select a 100K ohm resistor for R_c.

The compensation resistor works by applying the same voltage to the opposite input. Recall that these inputs are differential, meaning that the output will be zero when identical potentials are applied to the two inputs. This causes the voltage drop across the feedback/input resistor combination to be cancelled by the voltage drop across the compensation resistor R_c.

There are actually several causes of output offset potentials, and not all of them are dealt with so handily as the input bias current problem. Figures 14-2 and 14-3 show two methods for dealing with a multitude of offset problems, including some occasions where we want to *put in* an intentional offset potential.

The circuit of Fig. 14-2 shows a technique for use when the specific operational amplifier has *offset null* terminals. About half of the commercial operational amplifiers on the market provide some means of offset null. A potentiometer of usually 10K to 100K ohms is connected between the two offset null terminals on the operational amplifier. The wiper of the potentiometer is connected to the V_{ee} (V−) power supply. The output voltage is monitored, with the input voltage set to zero. Then adjust the potentiometer for zero volts output. In many cases, the potentiometer is actually adjusting the ratio of the two currents in the differential input transistors (I1 and I2 in Chapter 9). In other cases, the potentiometer will adjust the stage immediately following the input preamplifier.

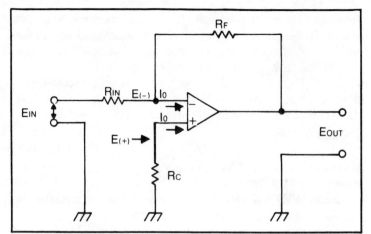

Fig. 14-1. Resistor compensation for input bias currents.

The method of Fig. 14-2 offers trimming over a relatively limited range on most operational amplifiers. In those devices which do not have offset null terminals, such as the 1458 device, this circuit is useless. An alternative circuit in Fig. 14-3 works with all operational amplifiers, regardless of whether offset null terminals are provided.

Figure 14-3 uses an offset null current input to the summing junction (point X) from the offset null potentiometer through R1. The overall output voltage of this circuit is given by the expression:

$$E_{out} = -\left[\frac{E_{in}R_f}{R_{in}} + \frac{E_A R_f}{R1}\right] \quad \textbf{Equation 14-2}$$

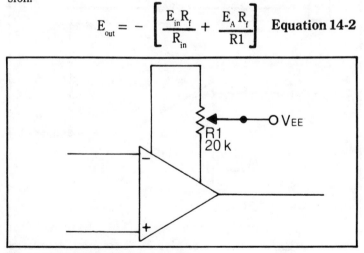

Fig. 14-2. Offset compensation techniques.

141

Once again, set input voltage E_{in} to zero and measure the output voltage. It should be zero in an ideal operational amplifier. In the real world, however, there will probably be some offset voltage present. Adjust E_A to cancel the offset—adjust the potentiometer R2 until E_{out} is zero when E_{in} is also zero. Note that potentiometer R2 is connected across the V_{cc} and V_{ee} power supplies, so it is capable of delivering either positive or negative values of E_A.

☐ **Example:**

A gain-of-100 operational amplifier circuit has a feedback resistor of 1 megohm. Find the value of E_A that will cancel a $-4V$ offset when R1 is 10K ohms.

Because $E_{in} = 0$, you must conclude that the output offset is cancelled when the following condition is met:

$$-4\text{volts} = (E_A)(1,000,000\text{ ohms})/(10,000\text{ ohms})$$

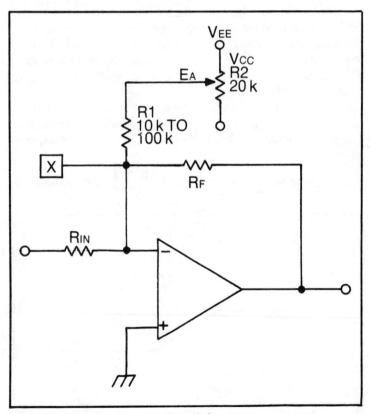

Fig. 14-3. Another offset compensation technique.

The value of E_A, then, is:

$$E_A = \frac{(-4 \text{ volts}) (10,000 \text{ ohms})}{(1,000,000 \text{ ohms})}$$

$$= -40,000/1,000,000 = 0.04 \text{ volts}$$

Note that the voltage required to null a large 4V offset is only 40 millivolts. This is a small voltage and is difficult to obtain with a simple potentiometer method as shown in Fig. 14-2. Of course, some higher resolution per turn of the potentiometer is possible by using a high-value resistor for R1. For example, the value of R1 in the above case had been increased to 100K ohms, a 0.4V potential would have been required. Similarly, if the value of R1 was 1 megohm, 4 volts would be required for E_A. But increasing the value of the circuit resistances causes other problems, to be dealt with in our section on noise in operational amplifier circuits.

We can, however, increase the resolution of the circuit by any of the circuits shown in Fig. 14-4. In the example of Fig. 14-4A we increase the resolution by placing series resistors in each arm of the potentiometer. If we can make the potentiometer 10 percent of the overall resistance (R1 = 0.1 (R1 + R2 + R3)), we have greatly improved the resolution of the trimmer circuit.

The same result is obtained in Fig. 14-4B by connecting low-value zener diodes (V_z is much less than V_{cc} or V_{ee}) between the ends of the potentiometer and ground. In most cases, the value of these zener potentials are equal to each other and will rarely be more than a few volts. In this type of circuit, the value of E_A will be a fraction of the difference in zener potentials, instead of $V_{cc} - V_{ee}$.

A combination circuit is shown in Fig. 14-4C. This circuit uses two trimmer potentiometers to perform offset nulling. A course potentiometer is the circuit of Fig. 14-3, while a fine control is formed using one of the other two, high resolution, circuits.

OPERATION FROM A SINGLE SUPPLY

The operational amplifier normally requires bipolar power supplies. One of the two power supplies, denoted V_{cc} or V+, must have a potential that is positive with respect to ground. How then can a bipolar operational amplifier be operated from a unipolar power supply? The answer—at least a limited solution—is shown in Fig. 14-5. This inverting follower circuit is similar to the circuits in Chapter 10, except for the power supply and noninverting input configuration. The V_{cc} terminal of the op amp is connected to a V+

voltage source, as is the usual practice. The V_{ee} power supply terminal, however, is grounded. To compensate for this, a voltage must be applied to the noninverting input. This is done by a voltage divider made from R1 and R2. The voltage applied to the noninverting input will be:

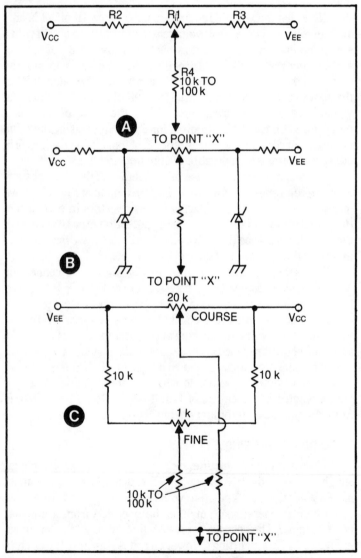

Fig. 14-4. Offset compensation techniques using one resistor (A), resistor-diode arrangements (B) and network (C).

$$E1 = \frac{(V+)(R2)}{(R1 + R2)}$$ **Equation 14-3**

If R1 = R2, the voltage at the noninverting input will be ½– (V+). But what does this do to the output voltage? The output voltage will have a constant ½(V+) DC component. In most cases, this type of circuit is used as an AC amplifier in mobile equipment or other applications where unipolar power supplies are a minority feature. A DC blocking capacitor in the output line will eliminate the DC component.

Another solution is to provide a DC-to-DC converter that will supply the V+ and V– power supply potentials for the operational amplifier. Most of these circuits are oscillators that drive a transformer. We could generate a 100-kHz AC signal and make the transformer part of the oscilloscope inductor. The signal at the output of the transformer is then rectified and filtered (100 kHz requires very little filtering) to make nearly pure DC. This technique is used extensively, and DC-to-DC converters are commercially available as circuit building blocks. Most will accept an input voltage of 6 to 28 volts and produce an output of either +12 or +15 volts DC at some low current (5 to 100 mA).

NOISE

The perfect operational amplifier is noise-free. High-quality commercial operational amplifiers produce so little noise as to be considered almost noise-free, but there is a certain residual noise that is dependent upon factors such as the circuit bandwidth and the external resistances of the op amp. These noises will mar the output signal, even when the operational amplifier, is itself, totally noiseless! This *terminal noise* is given by:

$$\overline{E}_n = \sqrt{4KT\,\Delta FR}$$ **Equation 14-4**

where $\overline{E}_n$ is the rms value of the noise, K is Boltmann's constant $(1.38 \times 10^{-23} J/°K)$, T is the temperature in degrees Kelvin (°K), ΔF is the bandwidth of the circuit, and R is the resistance that generates the noise signal.

You can see, then, that there is a normal input noise that is causes solely by the thermal action in the feedback and input resistances. For all practical purposes, the value of R in Equation 14-4 is the parallel combination of these resistances.

In high-gain circuits, this noise contribution can make things bad for weak signals. The K term cannot be changed, because it is a constant. Similarly, we cannot always affect the temperature of the circuit, because this may be an operating parameter with which we must live. The ΔF and R terms can be affected, however. In order to keep the noise low, then, we must use the lowest resistor values consistent with proper operation of the circuit, and a bandwidth that is no wider than necessary to transmit the input signal with good fidelity. In some cases, we might want to design the "front end" with low noise in mind, and then obtain the bulk of our gain in the stages that follow.

The bandwidth of an operational amplifier circuit is sometimes controlled by placing a capacitor in parallel with the feedback resistor, as shown in Fig. 14-5. The rolloff will begin at a frequency given approximately by:

$$f = \frac{1}{(6.28)R_f C} \qquad \textbf{Equation 14-5}$$

We can limit the noise amplitude in the circuit by selecting a value for the capacitor that will tend to roll off the gain of the circuit at

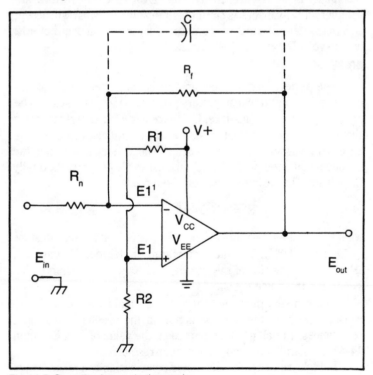

Fig. 14-5. Operation from a single-supply

some frequency slightly higher than the desired operating frequency. All complex (not sine) waveforms are actually a summation of sines and cosines that are harmonically related to each other. This is called the *Fourier series* of the waveform. There will be some component, a harmonic of the fundamental frequency, that will be the maximum frequency needed to pass the waveform with no apparent distortion. The value of F in Equation 14-5 should be at least this frequency.

THERMAL DRIFT

The basic operational amplifier is a DC amplifier, and as such will exhibit a certain thermal sensitivity. All transistors are basically temperature sensitive; in fact, the basic equation for base-emitter voltage contains a temperature term. Operational amplifier manufacturers will go to some length to stabilize their products, but all will have at least some temperature instability.

The thermal specification for operational amplifiers is given in terms of a potential change per unit of temperature change. A typical specification might be 1 to 5 $\mu V/°C$. This drift is for a steady-state test condition and may well be the least that can be expected under dynamic conditions.

At high-signal levels, we can all but ignore the thermal drift of the op amp. If the typical input signal is on the order of 100 to 1000 millivolts, then we don't worry too much about a 3 $\mu V/°C$ drift specification. But what about the situation where we are trying to amplify very weak signals? Some transducers and most biological signal sources have 1 mV as the upper amplitude. Many human brain-wave signals, for example, are in the 5 to 100 μV range. In this range, a 3 $\mu V/°C$ op amp drift becomes a major component of the offset and will be critical. Consider the human ECG signal as another example. It has an amplitude at the peak of 1 mV, with some important features having amplitudes of 0.1 mV. If we have an operational amplifier with a 5 $\mu V/°C$ drift property, and the temperature range changes by 20 degrees, the drift changes 100 μV or 0.1 mV: the artifact is 10 percent of the maximum peak signal amplitude and is equal to some features.

One method for improving the drift will be to insert a compensation resistor between the noninverting input and ground; the same technique as was used in a previous section to eliminate the bias-induced offset potential. The value of the compensation resistance, R_c, is the parallel combination of the feedback and input resistors.

147

OSCILLATION AND INSTABILITY

Several different factors will cause an operational amplifier to oscillate. Interestingly enough, these oscillations often occur at frequencies greatly in excess of the frequency response of the circuit connected to the op amp. Two important factors in creating unwanted oscillation are *positive feedback* via the DC power supply and *internal phase shift*.

The power supply for any amplifier should be a low impedance at all frequencies. The filter capacitors sometimes help, but be aware that electrolytic capacitors will not work well at high frequencies. Sometimes, designers will connect a small value disc ceramic capacitor in parallel with the electrolytics in order to reduce the impedance at radio frequencies. There is also sometimes a problem with the value of electrolytics. It may well be sufficient to cause filtering, especially if the supply is regulated, but may be a high impedance at some very low frequency. This may cause a type of oscillation similar to what is called *motorboating* in audio amplifiers.

The best solution to the problem of high power-supply impedance is to connect bypass capacitors from the V_{cc} and V_{ee} terminals to ground. These bypass capacitors should be disc or mica in most cases, or one of the special dielectrics now available. The value will be in the range 0.001 to 0.1 μF in most cases and as high as 1 μF in some circuits (Fig. 14-6). The only really important rule of thumb is to make sure that the capacitors are physically located as close as possible to the body of the op amp package. A few centimeters of printed-circuit track, or worse yet, hookup

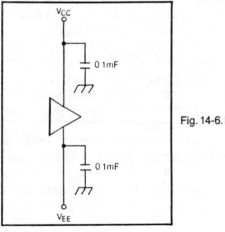

Fig. 14-6. Power-supply line bypassing.

148

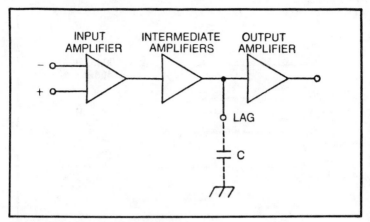

Fig. 14-7. Lag compensation (capacitor).

wire, can easily create a high impedance at some frequency where the operational amplifier wants to oscillate. Some devices, incidentally, are quite capable of producing oscillations in the upper-HF and lower-VHF regions.

Thus far, we have made things easier on ourselves and considered the operational amplifier as dealing with purely resistive circuits. In some frequency-compensated operational amplifiers, this is a reasonable assumption. But in most uncompensated models, this assumption can lead to disaster. There are several reactances in the circuit. One is the junction capacitance of the input transistors. Also, a capacitance is between the input terminal and the substrate of the chip. Also, ordinary stray capacitances are between the IC and its wiring and ground.

Regardless of the source of the capacitance, the end result is an RC phase shift. In the inverting amplifier configuration, a normal

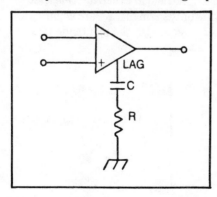

Fig. 14-8. RC lag compensation.

149

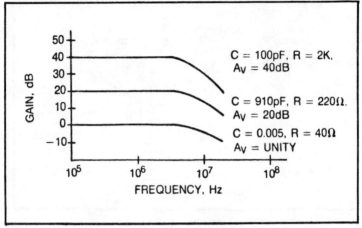

Fig. 14-9. Frequency response for values of R and C.

180-degree phase shift is between input and output, exactly as required for the degenerative feedback that is so much needed in op amp circuit design. Consider now the normal criteria for oscillation in an electronic circuit: 360-degree phase shift and gain of unity or more. If the internal phase shifts caused by the capacitances are capable of giving a 180-degree phase shift at some frequency at which the op amp has a gain of unity or more, a potential for oscillation exists. The additional phase shift is the 180 degrees from the inverter action.

Frequency compensation is the act of identifying the frequency at which the oscillation will occur and then reducing the gain of the operational amplifier at that frequency (and higher) to less than unity. There are two basic forms of compensation: *lag* and *lead*. *Input compensation* is another form.

Figure 14-7 shows the basic technique called *lag compensation*. The idea here seems a little brutish, in that we are shunting a relatively large-value capacitor to ground at the input side of the

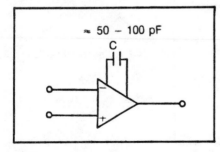

Fig. 14-10. Lead compensation.

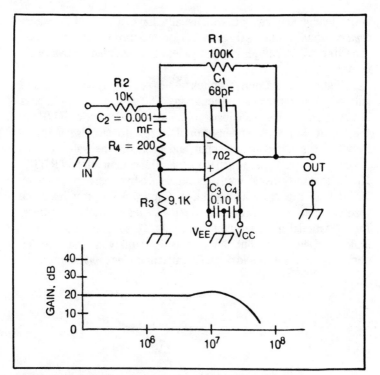

Fig. 14-11. Fully compensated operational amplifier circuit.

last amplifier in the stage (it's difficult to say, "the input side of the output amplifier"). The proper point for lag compensation is usually brought out to the op amp pins, and the terminal marked *lag* in the pinout chart.

An alternative lag technique is to connect an RC network to ground (Fig. 14-8) from the lag terminal. The idea here is to roll off the gain to unity at the oscillation frequency, but leave it relatively high, or at least *flat* at frequencies close to the breakpoint of the RC network ($f = 1/6.28RC$). This circuit allows us to custom tailor the frequency response of the circuit and is sometimes used to shape the frequency response quite apart from the oscillation problem. Figure 14-9 shows the effect of different values of resistance and capacitance as the lag terminal of a 702 operational amplifier device.

Lead compensation involves shifting the frequency at which oscillation occurs to some higher frequency. The idea is to make the frequency at which the problem occurs in the range *above* the unity gain frequency of the op amp. Operational amplifiers that

151

allow lead compensation will usually have a pair of *frequency compensation* terminals (Fig. 14-10) for this purpose. A small value capacitor (27 to 220 pF in most cases) is connected across these terminals.

Figure 14-11 shows an example of a fully compensated gain-of-10 operational amplifier stage. The gain of the amplifier is set by the ratio of the feedback to the input resistors, or R1/R2 in the circuit shown. A compensation (DC) resistor (R3) is connected between the noninverting input and ground. The value of this resistor is 9.1K ohms, which is the parallel combination of R1/R2. Notice that the power-supply terminals are bypassed to ground by two 0.1-μF capacitors (C3 and C4). Capacitor C1 provides some lead compensation. Notice the C2/R4 series RC network across the operational amplifier input terminals. This is sometimes called *input compensation*, but since it is essentialy a brand of lag compensation, we should properly call it *input lag compensation*.

Chapter 15
Amplifier Projects

In this chapter, some of what has thus far been discussed about operational amplifiers is applied. All projects have been built and tested by the author, and some of them are still in daily use by the client. You may wish to build these circuits as shown or modify them to your own purpose.

UNIVERSAL REAR END

Almost any serious amplifier project will require some form of rear end, or output stage (Fig. 15-1). Such an output stage should provide several functions: sensitivity control, position control, gain (if any additional gain is needed) and DC balance control. This latter control is used to cancel the effects of offset potentials that may have accumulated through the preceding changes. It is wasteful and bad engineering practice in most circuits to null each stage. A better method is to null some later stage, such as with the *DC balance control*. DC balance gets its name from the fact that the sensitivity control will cause a base-line shift in the output as it is rotated because it treats the input offset as a valid DC signal. The DC balance control is adjusted until the sensitivity control can be operated through its entire range without causing a significant base-line shift at the output.

Operational amplifier A1 is the input stage and is in the form of an inverting follower. If unit gain is required, make resistor R2 equal to R1. But should gain be required, make resistor R2 a value that will provide the required additional gain ($A_v = R2/R1$). The input stage could just as easily have been a noninverting amplifier, but in the original project, a phase inversion was required. Also, because an even number of inverting stages are in the "active" part of this circuit, phase reversal would not be obtained without an additional amplifier.

Adjustment Procedure

Set the R5 and R10 to midrange and R7 to maximum. The procedure is as follows:

☐ Ground point A

☐ Connect a DC voltmeter to point C

☐ Adjust R11 for 0 VDC ($\pm$ 10 mV) at point C.

☐ Connect a calibrated DC oscilloscope to point D.

☐ Adjust potentiometer R7 through its range from zero to maximum.

☐ Adjust R5 to cancel any base-line shift observed in the last step.

☐ Repeat the last two steps until there is no base-line shift as R7 is varied through its entire range. Note that high values of R2 will make this adjustment touchy—or impossible! Note that the DC balance can also correct for differences in any range switches that are used in stages preceding this rear-end amplifier.

"UNIVERSAL" PREAMPLIFIER

Figure 15-2 shows the circuit diagram for a "universal" (not really, but it does have wide application) preamplifier that is useful mostly in scientific and engineering instrumentation applications. The rear end, or output, is little more than a modification of the circuit shown in Fig. 15-1. In fact, the earlier circuit could be easily

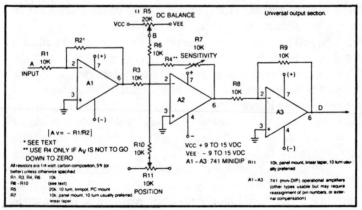

Fig 15-1. Universal output section.

Table 15-1. Parts List for the Instrumentation Amplifier.

Resistors are 1/4 watt, 5 % or better, unless otherwise noted.

R1	2.2k
R2, R3	39k
R4, R5	133k, 1%
R6	1 meg
R7	820k
R8	250k, 10-turn, trimpot, PC mount
R9, R10	47k
R11	100k
R12	150k
R13	10k or 100k
R14	100k, screwdriver adjust, miniature, panel mount potentiometer
R15, R17	10k
R16	20k, 10 turn, trimpot, PC mount
C1, C2	250 μF, 16 WVDC, or better
S1	DPST or DPDT miniature toggle
S2	SP3T rotary
J1	black banana jack (not 5-way binding post)
J2, J3	Grounded type banana jack
J4	Red banana jack, same as J1 except for color
J5	Phone jack or whatever matches your equipment.
J6	Amphenol 126-197 (mating 126-196 also needed)
A1, A2	RCA CA3140 operational amplifiers (specify DIL-pak)
A3, A4	MC1458 dual operational amplifier

4-mini-DIP IC sockets
Small piece of Vector 3677-2DP perfboard
Pomona 2901 bluebox

substituted for stages A3B, A4A and A4B. The important part of the circuit is the preamplifier, or front end. This portion of the amplifier consists of A1, A2 and A3A, which form an instrumentation amplifier. The two input amplifiers (A1/A2) form a differential pair and must be of high input impedance design. It is worthwhile using a biMOS or biFET operational amplifier in these slots. In fact, it would be wise to use a dual biFET or biMOS device in this slot, so that the two will thermally track one another. The RCA CA3240 is a dual CA3140 device, so it should probably work nicely.

Adjustment Procedure

Set R8 and R16 to midrange, R14 to maximum and S2 to position 3. The procedure is as follows:

☐ Short together J1 and J4.

☐ Connect an oscilloscope or DC voltmeter to J5.

☐ Connect a DC voltmeter to point A.

☐ Adjust R16 for 0 V DC (±10 mV) at point A.

☐ Run R14 through its range from the maximum down to minimum and then back to maximum. Note whether any baseline shift showed up on the oscilloscope.

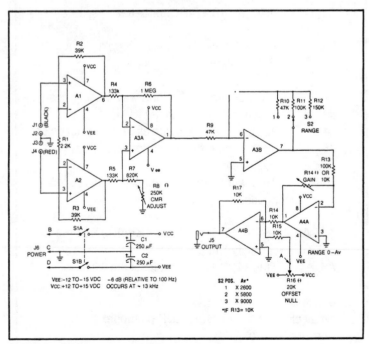

Fig. 15-2. Instrumentation amplifier.

156

☐ Adjust R8 for zero shift of base line when again adjusting R16 for 0 VDC at point A. This step may have to be repeated several times at increasingly more sensitive settings of the oscilloscope so that the best null can be found.

☐ Adjust R16 for 0 VDC (±10 mV) at J5.

☐ Remove the short from J1 and J4.

☐ Ground J4.

☐ Connect a signal between J1 and ground. This signal should have an amplitude under either 10 mV or 1 mV, depending upon which value you selected for R13.

☐ Check for distortionless amplification.

☐ Disconnect the signal source and unground J4.

☐ Ground J1 and connect the signal source between J4 and ground.

☐ Again check for distortionless amplification.

☐ Use the amplifier—it is ready.

Experimenters in the life sciences and physical sciences and certain engineers often use Wheatstone bridge transducers or other resistive transducers connected into a Wheatstone bridge circuit. These transducers will generate a voltage analog of the parameter that strains the transducer. Unfortunately, it is the nature of precision transducers of this genre to produce very weak output signals.

Usually, the transducer is connected to a carrier amplifier or some amplifier with a fixed DC reference source. Unfortunately, these amplifiers tend to be expensive. The transducer preamplifier shown in Figs. 15-3 and 15-4 were built for a physiologist who was trying to operate the transducer directly into a Tektronix oscilloscope. The scope in his laboratory had a 200 μV/cm position on the vertical deflection range switch, so this seemed at first blush to be a reasonable alternative. But the signal, which was a force-displacement transducer (Fig. 15-4) output, was noisy. A large, 60-Hz interference signal was present on the output. The preamplifier was designed to amplify the transducer signal before applying it to the shielded line to the oscilloscope input. This tactic is standard practice for overcoming signal-to-noise ratio problems. A connector that mates with the connector on the Grass force-displacement transducer was mounted on one end of the installation box, and an Amphenol series-126 connector for power and signal was mounted to the other end.

The operational amplifier is a model CA3140 or CA3160 by RCA. This device was selected for its low cost and low offset

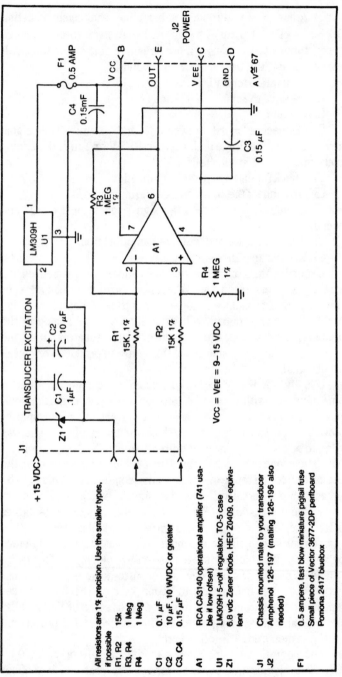

Fig. 15-3. Transducer preamplifier.

All resistors are 1% precision. Use the smaller types, if possible

R1, R2	15k
R3, R4	1 Meg
R4	1 Meg

C1	0.1 μF
C2	10 μF, 10 WVDC or greater
C3, C4	0.15 μF

A1   RCA CA3140 operational amplifier (741 usable if low offset)

U1   LM309H 5-volt regulator, TO-5 case

Z1   6.8 vdc Zener diode. HEP Z0409, or equivalent

J1   Chassis mounted mate to your transducer
J2   Amphenol 126-197 (mating 126-196 also needed)

F1   0.5 ampere, fast blow miniature pigtail fuse
Small piece of Vector 3677-2DP perfboard
Pomona 2417 bluebox

Fig. 15-4. Transducer preamplifier of Fig. 15-3.

potential. It is, however, quite possible to hand select a 741 to work in this application. The gain of this circuit is about 67. The unusual figure for gain was selected for a very good reason: I had 15K ohm and 1 megohm precision resistors on the bench. You might want to select some reasonable gain, like X10, X50 or X100, but only something of 50 or more was required in this case.

The Wheatstone bridge transducer requires a DC excitation source. It is necessary to heed the manufacturer's recommendation as to the highest permissible excitation voltage; otherwise, you might damage the transducer. In this case, the transducer would take as much as 10 volts, so the 5 volts selected was well within the correct range. The LM309H regulator was mounted inside the case with the operational amplifier. The H-series LM-309 was in a TO-5 transistor case, and it will source up to 100 mA of current at +5 volts output. The V+ for the LM-309H should be anywhere from +7.5 to +15 volts, although using the lowest possible DC voltage is recommended.

The zener diode across the +5V line is a means of protecting the transducer should a catastrophic short circuit develop in the LM-309H. If we were using 12 volts as the V+, for example, a short would place 12 volts across the transducer, causing a possible burnout. The value of the zener potential should be less than the burn-out voltage of the transducer or less than V+, which ever is more appropriate. In this case, a 6.8V zener was selected.

Chapter 16
Analog Active Filters

A *filter* is a circuit that will pass certain specified frequencies and reject all others. Passive components are often used to make filters; RL, RC, and RLC networks have frequency sensitive characteristics because of the frequency terms in the expressions for inductive and capacitive reactance. But passive filters are difficult to implement in many situations, especially where inductance values become quite large. An active filter uses an amplifier device, with a frequency sensitive (usually RC) network in the feedback loop.

There are several basic types of filter circuits, classified by their respective pass or rejection bands. The different types are *low-pass, high-pass, bandpass*, and *notch*. The frequency response characteristics for these filters are shown in Fig. 16-1.

LOW-PASS FILTERS

A low-pass filter is designed to pass all frequencies between DC and some specified *cutoff frequency* (f_c). A 500-Hz filter, then, will pass all signals between DC and 500 Hz. A perfect or ideal filter will abruptly cut off at f_c. But real filters are not quite so tidy; they will gradually attenuate frequencies in the low-pass case higher than f_c. This action forms a *skirt* in the frequency response property shown in Fig. 16-1A. From the cutoff frequency, the response drops in a nearly linear manner, until, at some frequency greater than f_c, the response reaches zero.

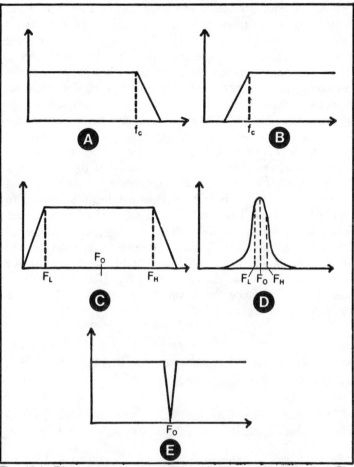

Fig. 16-1. Filter response characteristics: Low-pass (A), high-pass (B), bandpass (C), high-Q bandpass (D) and notch (E).

One means for representing the quality of a filter is in the steepness of the slope. The closer the filter is to the ideal, the more perfect the attenuation of frequencies outside the passband is. This is measured as the *slope* of the frequency response skirt outside the passband. Filter rolloff is usually specified in terms in *decibels per decade* (10:1 frequency change), or *decibels per octave* (2:1 frequency range). Note that the dB/octave specification will result in a number different that dB/decade. If two filters have the same number but used different ranges to obtain the dB figure, then the filter specified in terms of decibels per octave is the steeper slope. Note that −6dB/octave equals 20 dB/decade.

A simple RC low-pass filter is shown in Fig. 16-2. This circuit consists of a resistance in series with the signal line, and a capacitor shunted across the load. The resistor has a constant frequency response, but the capacitor has a reactance that is inversely proportional to the frequency. This means that the reactance will be lower as the frequency increases. If we view this circuit as a voltage divider $R1/(R1 + X_{C1})$, it is easy to see how the output voltage, for a constant input amplitude, will decrease as the frequency increases. The frequency response for this type of circuit is shown in Fig. 16-3. The response will be essentially flat at frequencies well below f_c. At those frequencies, the reactance of C1 will be high enough to have negligible effect on the output vottage. But at f_c, the output response will begin to drop more rapidly than before. The slope of the rolloff will be -6 dB/octave. Such a filter is known as a first-order filter. Higher order filters will have a faster rolloff rate. A second-order filter, for example, will have a rolloff of -12 dB/octave. Higher order circuits can be made by cascading first-order circuits, such as Fig. 16-2. Two sections cascaded will produce the second-order response. The cutoff frequency can be approximated by:

$$f_c = \frac{1}{2 \pi RC}$$

where f_c is the cutoff frequency in hertz, R is the resistance in ohms, and C is the capacitance in farads.

□ **Example:**

Calculate the cutoff frequency of a single-order RC low-pass filter in which $R = 2.2K$ ohms and $C = 0.001 \mu F$.

$$f_c = \frac{1}{2 \pi RC}$$

$$= \frac{1}{(2)(3.13)(2.2 \times 10^3 \text{ ohms})(1 \times 10^{-9} \text{F})}$$

$$= 1/(1.38 \times 10^{-5}) = 72,380 \text{ Hz}$$

HIGH-PASS FILTERS

Returning to Fig. 16-1, examine the response of the high-pass filter in Fig. 16-1B. A high-pass filter is the inverse of the low-pass response: It *attenuates* all frequencies between DC and the cutoff frequency. All frequencies higher than f_c will be passed. We can make an RC single-order high-pass filter by reversing the roles of the resistor and capacitor of the low-pass design (Fig. 16-2). If the capacitor is placed in series with the load, and the resistor is shunted across the load, a high-pass response is obtained.

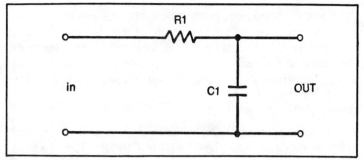

Fig. 16-2. RC low-pass filter.

You may have recognized the low-pass filter circuit as the same circuit used to *integrate* analog signals. Similarly, the high-pass version is the same circuit as a differentiator. In general, all integrator circuits are low-pass filters, and differentiators are high-pass filters. The reverse is not always true, however. We cannot assume that all low-pass filters will integrate or that all high-pass filters will differentiate. In those cases, the time constant R x C is important. An integrator requires a time constant that is long compared with the period of the applied signal, while a differentiator works only if the time constant is short compared with the period of the applied signal.

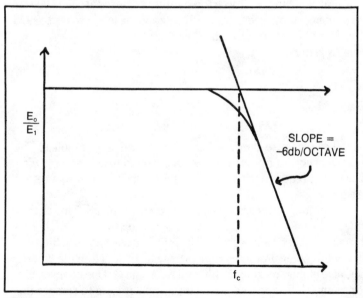

Fig. 16-3. Response curve of the high-pass filter.

Some electronic equipment manufacturer's service manuals sometimes erroneously list circuits incorrectly because of these facts. One medical equipment manual, for example, calls the integrator used in the heart rate meter a "low-pass filter." Yet, in reality, the circuit is the integrator section of an ordinary pulse tachometer circuit.

BANDPASS FILTERS

A bandpass filter is a combination of the low-pass and high-pass concepts (see Figs. 16-1C and 16-1D). The bandpass filter will pass only those frequencies that are above the low cutoff frequency and below the high cutoff frequency.

One measure of the quality of a bandpass filter is the *Q factor*, which is related to the center band frequency (f_o in Figs. 16-1C and 16-1D) and the bandwidth ($f_h - f_L$). Expressed as an equation:

$$Q = F_o/(BW)$$

or

$$Q = f_o/(f_h - f_L)$$

where Q is a dimensionless quality factor, f_o is the center band frequency, f_h is the high cutoff frequency, f_L is the low cutoff frequency, and BW is the candwidth ($f_h - f_L$).

☐ **Example:**

Calculate the quality factor of a filter centered around 1000 Hz if the upper cutoff frequency is 1075 Hz and the low cutoff is the 925 Hz. Also calculate bandwidth.

Quality factor—

$$Q = f_o/(f_h - f_L)$$
$$= (1000)/(1075\text{-}925)$$

Bandwidth—
$$= (1000)/(150) = 6.7$$

$$BW = f_h - f_L = 1075\text{-}925 = 150\,Hz$$

The response shown in Fig. 16-1D is a high-Q response. The high-Q bandpass filter is sometimes referred to as a *peaking filter* because it tends to pass only those frequencies close to a center frequency.

The notch filter response is shown in Fig. 16-1E. This class of filter will pass all frequencies, except those clustered about a certain specified frequency, f_o. The principal use for notch filters is to remove some unwanted frequency without affecting the others. Because of this, most notch filters have a high Q. One common use for the notch filter in electronic instrumentation is to remove the 60-Hz interference from AC power wiring or to remove the 120-Hz ripple component that might be on the power-supply output voltage. There are three basic forms of notch filter (RC, LC, and RLC), but the one that is of most importance to us is the RC network shown in Fig. 16-4. This circuit is sometimes called a *T-notch filter,* or a *twin-T filter*.

Either points A or B can be used as input or output terminals for this network, so from this point of view, the circuit is symmetrical. Point C, however, is the signal common, so it will often be grounded. The notch frequency is given by the first equation of the chapter, in which f_c is the notch frequency. With practical components, the notch depth will be in the 25 to 55 dB range. The actual depth obtained in any given instance will depend upon circuit layout and, most importantly, the nearness of the component values to the calculated values. To obtain the deepest notch requires proper layout (isolation of the components from each other to prevent coupling) and a closeness of the component tolerances. Unfortunately, standard component values do not often match the calculated component values, so exact values might be difficult to obtain. Also, there is great variation in any group of electronic components of the actual component values. All electronic components have a certain tolerance in their values. If you use 10 percent tolerance resistors, then expect results to be somewhat less exciting than could be obtained using 0.05 percent components. In some applications, it would be a good idea to hand sort capacitors and resistors, using precision instruments to locate those that most nearly meet the required specification. It is also necessary to make sure that the capacitor has a low temperature coefficient or the value will change with temperature, thereby degrading the notch.

Figure 16-5 shows the first *active* filter which we will consider. This is an elementary notch filter based on an operational amplifier and a twin-T network. The amplifier is connected with a heavy dose of negative feedback to the inverting terminal. The common point of the twin-T network is also connected to this point. The notch frequency is found from the same equation just mentioned, while the Q of the notch filter is given by:

$$Q = R_f/2R$$
$$Q = C/C_b$$

☐ **Example:**
   Design a 60-Hz notch filter with a Q of 2.
First, select a trial value for the capacitor. It is usually best to

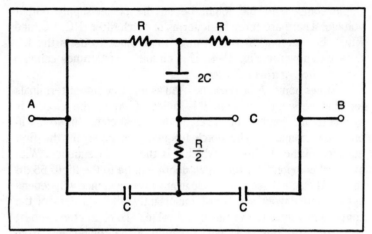

Fig. 16-4. Twin-T network.

select the value of C from a standard table, and then find an appropriate R value. Let us select $0.01\,\mu F$ as a trial.

$$R = 1/2\pi Cf_o$$
$$= 1/(2)(3.14)(1 \times 10^{-8}f)(60\,Hz)$$
$$= 2/(3.77 \times 10^{-6}) = 266K\,ohms$$

A chart of precision resistor values reveals that 267K ohms is a standard value and is within 1 percent of the calculated value. By hand selecting from a group of 267K ohm resistors, a 266K ohm, 1 percent resistor is possible. The value of R/2 will be 133K ohms, also a standard value.

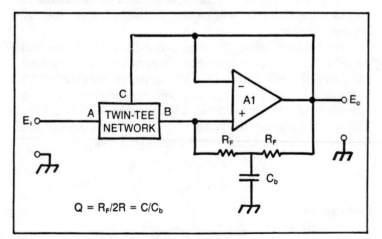

$$Q = R_F/2R = C/C_b$$

Fig. 16-5. Twin-T active notch filter.

Table 16-1. Notch Filter Values Centered on 60 Hz.

Component	Q = 2	Q = 4
R	267 K	68.1 K
R/2	133 K	34.05 K
R_f	1.06 meg	270 K
C	0.01 μF	0.039 μf
2C	0.02 μF	0.078 μF
C_b	0.005 μF	0.01 μF

Now select the feedback resistor, R_f. For a Q of 2, this resistor should be R_f = 2QR = 4R = (4)(266K) = 1.06 megohms (also obtainable in precision resistor form). The capacitor is 0.01 μF, so for a Q of 2 filter, the bypass capacitor should be C/2, or 0.005 μF. The values are then: R = 267K ohms, R/2 = 133K ohms, C = 0.01 μF, and C_b = 0.005 μF. Table 16-1 shows the values for Q of 2 and Q of 4 notch filters centered on 60 hertz.

The Q of the notch filter can be amplified by using a circuit such as Fig. 16-6. This circuit is the same as the previous circuit, with one additional amplifier. We can reduce the feedback component applied to the common terminal of the twin-tee network, causing an increase in Q factor. But this circuit must be used with care. If the signal is too great, then no increase in Q is obtained. But if it is too low, then the Q will go so high that *oscillation* may result.

FILTER HALF-POWER POINTS

When dealing with circuits such as the bandpass amplifier, we often refer to the *–6 dB point* or the *half-power point*. These points are the same and are merely different expressions of the same idea. In filters, the cutoff frequencies occur at the half-power points in the frequency response spectrum.

But isn't half-power represented by -3 dB, not -6 dB? Yes, but power is difficult to measure in any event. In small signal circuits where filters are used, it is even more difficult. *Voltage* is a lot easier to measure in precision. Let's calculate the half-power in terms of voltage (we can assume that the resistance is constant). Power is expressed by E^2/R, and at the center frequency, by (see Fig. 16-7):

$$P = E_o^2/R_L$$

If we assume symmetry in the response curve, then the power at the half-power frequencies f_h and f_L is

$$P = \tfrac{1}{2}(E_o)^2/R_L$$

or
$$P = (E1)^2/R_L$$

so
$$\frac{(E_0{}^2)}{2R_L} = \frac{(E1)^2}{R}$$

from which, we can calculate:
$$\frac{(E_0)^2}{2} = (E1)^2$$
$$(E_0)^2 = 2(E1)^2$$
$$E_0 = (2)^2 E1$$
$$E_0 = (1.414)E1$$

so
$$E1 = E_0/1.414 = 0.707E_0$$

To measure the half-power point, then, find the points where the output voltage has fallen off to 0.707 of the output voltage at f_0.

The formulas in this book contain the expression $2\pi f$. In many electronics textbooks, this, expression is represented by the lower case Greek letter ω (omega):

$$\omega = 2\pi f$$

OBTAINING HIGHER ORDER FILTERS

Earlier we decided that one way to obtain higher order filters is to cascade several RC filters. The attenuation of the high frequencies, as expressed by the rolloff factor in dB/octave or dB/decade, will double for each section added. The general expression is N(−6 dB/octave), in which N is the number of sections, but

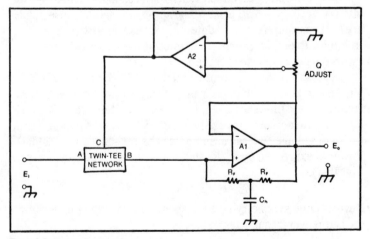

Fig. 16-6. Increasing the Q.

there is a loss to frequencies other than those outside of the passband. All RC filters have a certain *insertion loss*. Cascading several sections merely serves to increase the loss to in-band frequencies. One solution to this problem is to buffer each section from the others with an operational amplifier that has enough gain to overcome the losses. Similarly, the operational amplifier could overcome the loss of two or three sections, thereby reducing the number of operational amplifiers needed. However, this is not an ideal solution. A better solution is to use an active filter, but the topic of active filters is too complicated for the single-chapter treatment given here. We will consider only certain of the filter types, namely the unity gain, maximally flat bandpass, second-order types. We will also simplify the task by assuming from the outset that the Q of the filter will be $(2)^{1/2}/2$ and that certain resistor and capacitor ratios are maintained. These are specified and discussed as each case arises.

Figure 16-8 shows the general form of an active filter circuit. This circuit consists of an operational amplifier, connected in the unity-gain, noninverting configuration, and a network of impedances. These impedances are shown here in their general form,

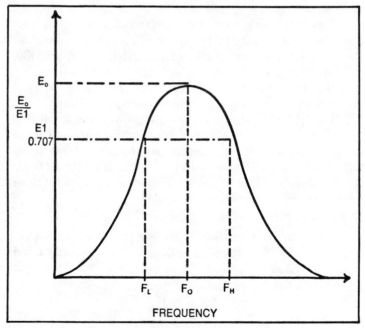

Fig. 16-7. Response measurement points.

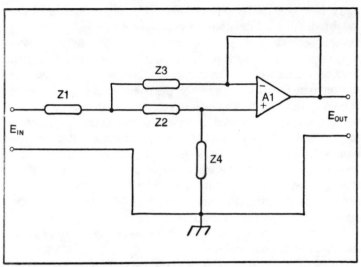

Fig. 16-8. VCVS filter general circuit.

because the specific impedances used at any given slot might be a resistance, a capacitive reactance, or an inductive reactance.

The operational amplifier can be a lowly 741 device in most applications, but the operation of the circuit will be improved significantly by using a premium operational amplifier. A good choice would be a device that is capable of a high input impedance. Recommended are the LF156 series, and the RCA CA3130, CA3140, and CA3160 devices. All four of these operational amplifiers have extremely high imput impedance values, and will therefore not load the impedance circuit making up the filter.

This type of filter is second-order, so will offer a rolloff of approximately −12 dB/octave if proper choice to component values are made. Those components will be capacitors or resistors, in most cases. Their values are so critical to proper operation of the circuit that either hand selection or obtaining very close tolerance precision types is recommended.

LOW-PASS CONFIGURATION

This circuit configuration for the low-pass filter is shown in Fig. 16-9. In this case, impedances Z1 and Z2 are resistances, and impedances Z3 and Z4 are capacitive reactances. We will make the job somewhat easier by specifying that:

$$C2 = 2C1$$

and

$$R1 = R2$$

One reason for making these specifications is that it will simplify the mathematics (filter math can get quite involved), allowing us to dispense with the complex notation of the *Laplace transform*. If the relationships of these two equations are met, then the low-pass cutoff frequency for the circuit in Fig. 16-9 will be expressed by:

$$f_c = \frac{1}{2\pi R_2 (C_1 C_2)^{1/2}}$$

where f_c is the cutoff frequency in hertz, R_2 is the resistance of R2 in ohms, C_1 is the capacitance of C1 in farads, and C_2 is the capacitance of C2 in farads.

☐ **Example:**

Find the resonant frequency of a low-pass filter (Fig. 16-9) in which the value of R2 is 100K ohms and the value of C1 is $0.005\,\mu F$. ($C2 = 2C1 = (2)(0.005\,uF)$, or $0.01\,\mu F$):

$$f_c = \frac{1}{(2)(3.14)(10^5\,\text{ohms})(5 \times 10^{-9}\text{f})(10^{-8}\text{f}))^{1/2}}$$
$$= 1/(4.44 \times 10^{-3}) = 225.2\,\text{Hz}$$

In designing an actual filter, you would probably either select C1 by trial and error or consult a chart in a filter design textbook. With the trial value of C1, the calculated value of C2, and the frequency specified for f_c, R1 and R2 could then be calculated by rearranging the last equation to solve for R2.

HIGH-PASS CONFIGURATION

The high-pass configuration for this class of filter is shown in Fig. 16-10. The high-pass response is the inverse of the low-pass response, so, as one might expect, the roles of the general

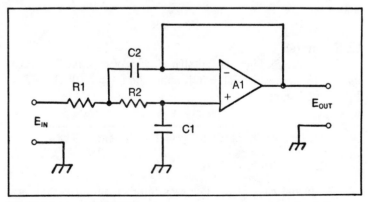

Fig. 16-9. Low-pass VCVS filter.

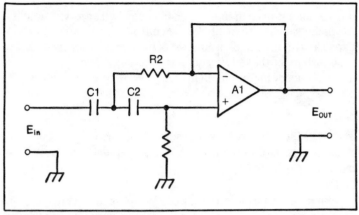

Fig. 16-10. High-pass VCVS filter.

impedances are reversed in this circuit compared with the
low-pass circuit of Fig. 16-9. In this circuit, impedances Z1 and Z2
are capacitive reactances, and Z3/Z4 are resistances. The cutoff
frequency is expressed by an equation similar to the low-pass case:

$$f_c = \frac{1}{2\pi C_1 (R_1 R_2)^{1/2}}$$

where f_c is the cutoff frequency of the high-pass filter, C_1 is the
capacitance of C1 in farads, R_1 is the resistance of R1 in ohms, and
R_2 is the resistance of r_2 in ohms.

This equation is valid only under certain specified conditions.
You must assume that the following resistance and capacitance
ratios are maintained:

$$C1 = C2$$

and

$$R2 = \frac{1}{2}R1$$

☐ **Example:**

Calculate the high-pass cutoff frequency of an active filter such
as Fig. 16-10 in which R1 = 82K ohms and C1 = 0.01 μF. (R2 =
½R1 = 41K ohms).

$$f_c = \frac{1}{(2)(3.14)((8.2 \times 10^4 \text{ ohms}) (4.1 \times 10^4 \text{ ohms})^{1/2} (1 \times 10^{-8} \text{ f})}$$

$$= 1/(3.64 \times 10^{-3}) = 275 \,\text{Hz}$$

As in the case of the low-pass filter circuit, it is standard
practice to select a value for the capacitor, either by trial and error

or through the use of a filter design chart, and then calculate the value of the resistor by solving for $(R_1 R_2)^{1/2}$. From there, the values of R1 and R2 could be selected. The reason for this procedure is that capacitors are available in fewer values than resistors. In addition, there are few ways to trim a capacitance, especially in the range of values normal to active filter design. Even in ranges where trimmer capacitors are available, they are somewhat more clumsy than trimmer potentiometers. Hence, it is usually better to select the capacitance to meet a standard value and then calculate the value of the required resistance. In most cases, a precision resistor can be found that will have the correct value. Consulting a listing of precision resistor values will reveal that there are more values listed than are available in the nonprecision lines of the same manufacturers.

A bandpass filter will result if you cascade a low-pass filter and a high-pass filter. The values of f_c for each filter stage will be selected such that they correspond to f_L and f_h for the desired bandpass response.

A low-pass filter will be designed with a cutoff frequency at the high cutoff desired in the bandpass filter. Similarly, the high-pass filter will be designed with a cutoff frequency at the low end limit in the bandpass response. This technique is used by a lot of designers and is considered quite valid. But in the next section, a circuit behaves as a bandpass filter but uses only one operational amplifier and impedance network to obtain the desired -12 dB/octave, second-order response characteristic.

MULTIPLE FEEDBACK PATH FILTER CIRCUITS

The general form for the multiple feedback path filter (MFPF) is shown in Fig. 16-11. This particular circuit is a little more difficult to solve than the voltage-controlled-voltage-source (VCVS) filters of the last section. But, in general, the MFPF filter circuit will yield superior results. The general form of the transfer equation is:

$$\frac{E_{out}}{E_{in}} = \frac{1}{(1/Z_5)((1/Z_1)+(1/Z_2)+(1/Z_3+(1/Z_4))+((1/Z_3)(1/Z_4))}$$

This equation is valid provided only that the operational amplifier has an extremely high open-loop voltage gain. In the correct expression, there are several terms involving $1/A_{vb}$, which is the open-loop gain at frequencies inside the bandpass limits. If the operational amplifier gain is high, then these terms will

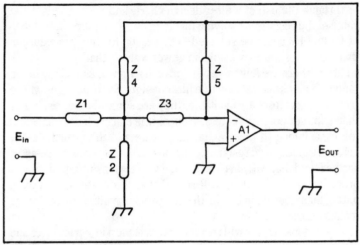

Fig. 16-11. MGPG general circuit.

approach zero, and the total expression reevaluates to that given in the equation.

Low-Pass MFPF Filter

Figure 16-12 shows the circuit for the low-pass version of the general MFPF design. Generalized impedances Z1, Z3, and Z4 are resistances in the low-pass version. Impedances Z2 and Z5 are capacitive reactances. In this case,

$$f_c = \frac{1}{2\pi (R_2 R_3 C_1 C_2)^{1/2}}$$

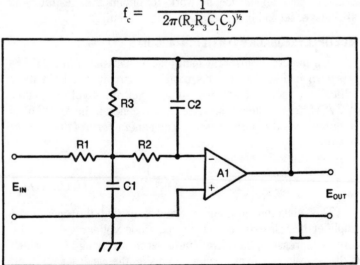

Fig. 16-12. Low-pass MFPF circuit.

174

or in the more commonly encountered form:

$$f_c = \frac{1}{2\pi\sqrt{R_2 R_3 C_1 C_2}}$$

where f_c is the cutoff frequency in hertz, R_2 is the resistance of R2 in ohms, R_3 is the resistance of R3 in ohms, C_1 is the capacitance of C1 in farads, and C_2 is the capacitance of C2 in farads.

Take advantage of the assumption that the Q of these circuits is set to $(2)^{1/2}/2$, or approximately 0.707. The gain inside of the passband is expressed by:

$$A_{vb} = R_3/R_1$$

If the ratio of the capacitance, C_1/C_2, is set to be a constant, denoted by k, then set:

$$k = 40^2(A_{vb} + 1)$$

then you are reasonably justified in claiming that:

$$C_1 = kC_2$$

and therefore:

$$C_1 = C_2(40^2(A_{vb} + 1))$$

By substituting in the initial assumption regarding the value of Q, $Q = (2)^{1/2}/2$, then:

$$C_1 = C_2(4((2)^{1/2}/2)^2(A_{vb} + 1))$$
$$= 2C_2(A_{vb} + 1) \qquad \textbf{Equation 16-1}$$

The last equation gives the value for C1, given an assumed value for C2. The value of resistance R2 is expressed by:

$$R_2 = \frac{1}{\omega_o^2 C_1^2 R_3 k}$$

$$= \frac{1}{8\pi^2 f_c C_1^2 R_3 (A_{vb} + 1)} \qquad \textbf{Equation 16-2}$$

The values for the other components, given the initial assumptions, are:

$$R_1 = R_3/A_{vb} \qquad \textbf{Equation 16-3}$$

$$R_3 = \tfrac{1}{4}\pi f_c Q C_2 \qquad \textbf{Equation 16-4}$$

175

$$R_3 = \frac{1}{2}(2)^{\frac{1}{2}}f_c C_2 \qquad \textbf{Equation 16-5}$$

The protocol for the selection of component values assumes a given sequence of selection, which is:

☐ Set C_2 to a convenient value.
☐ Calculate the value of C1 using Equation 16-1.
☐ Compute R_3 using either Equation 16-4 or 16-5.
☐ Compute the value of R1 using Equation 16-3.
☐ Compute the value of R2 using Equation 16-2.

☐ **Example:**
Design a 1-kHz second-order low-pass filter of the MFPF type (Fig. 18-12). Assume that $Q = (2)^{\frac{1}{2}}/2$ and that the gain inside the passband is unity ($A_{vb} = 1$).

Select C2 = 0.001 μF.

Calculate C1—
$$\begin{aligned} C1 &= 2C_2(A_{vb} + 1) \\ &= (2)(0.001\,\mu F)(1+1) \\ &= (4)(0.001\,\mu F) = 0.004\,\mu F \end{aligned}$$

Calculate R3—
$$\begin{aligned} R3 &= 1/(2)(2)^{\frac{1}{2}}f_c C_2 \\ &= 1/(2)(1.414)(10^3 Hz)(10^{-9} F) \\ &= 1/(2.828 \times 10^{-6}) = 353.6K \text{ ohms} \end{aligned}$$

Calculate R1—
$$\begin{aligned} R1 &= R3/A_{vb} \\ &= R3/(1) \\ &= R3 = 353.6K \text{ ohms} \end{aligned}$$

Calculate R2—
$$\begin{aligned} R2 &= 1/(8)(3.14)^2(f_c)^2(C1)^2(R_3)(A_{vb}+1) \\ &= 1/(8)(3.14)^2(10^3)^2(4 \times 10^{-9})^2(1+1) \\ &= 1/(8)(9.859)(10^6)(1.6 \times 10^{-17})(3.536 \times 10^5)(2) \\ &= 1/(8.925 \times 10^{-4}) \text{ ohms} = 1.1121K \text{ ohms} \end{aligned}$$

The first two steps in this procedure might have to be repeated several times before a correct set of values is obtained for the capacitor values. It is, for the same reasons as presented earlier, to select capacitor values that are standard values. Similarly, it may be then necessary to cut and try the whole procedure several times to obtain values for all components that are close to standard values.

High-Pass Filters

The high-pass MFPF filter is shown in Fig. 16-13. Once again, the fact that this circuit is the inverse of the low-pass design is reflected in the fact that the components have switched positions: Z1, Z3, and Z4 have become capacitors C1, C2, and C3, respectively: and impedances Z2 and Z5 have become resistances R1 and R2, respectively. The gain in this circuit is set by the capacitor ratio:

$$A_{vb} = C1/C3 \qquad \text{Equation 16-6}$$

and if a relatively flat passband is desired specify:

$$C1 = C2$$

The cuttoff frequency, f_c, is given by the expession:

$$f_c = \frac{1}{2 \pi \sqrt{R_1 R_2 C_2 C_2}}$$

Resistors R1 and R2 are selected from the following expressions:

$$R1 = \frac{1}{2 \pi f_c Q C_1 (2A_{vb} + 1)} \qquad \text{Equation 16-7}$$

$$R2 = \frac{Q(2 A_{vb} + 1)}{2 \pi f_c C_1} \qquad \text{Equation 16-8}$$

As in the previous case, the design is made a little easier if you follow a given sequence in your calculations: which is:

☐ Select C1 arbitrarily (make a good guess).

☐ Calculate C2. Set it equal to C1 in this case, given the assumptions.

☐ Calculate C3 using Equation 16-6, rearranged to solve for C3.

☐ Calculate R1 using Equation 16-7.

☐ Calculate R2 using Equation 16-8.

☐ **Example:**

Design a maximally flat (C1=C2) second-order, high-pass filter with a cutoff frequency of 100 Hz. Assume that the passband gain is unit (A_{vb}=1).

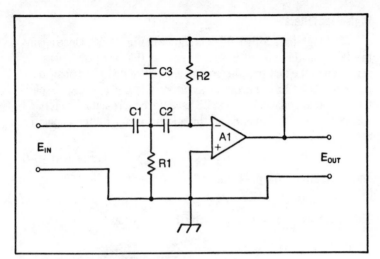

Fig. 16-13. High-pass MFPF circuit.

Set C1 = 0.01 μF.
  C1 = C2, so C2 = 0.01 μF.
Solve Equation 18-6 for C3–

$$C3 = C1/A_{vb}$$
$$= (0.01 \mu F)/1$$
$$= 0.01 \mu F$$

Calculate R1–

$$R1 = \frac{1}{2} \pi f_c Q C_1 (2A_{vb} + 1)$$
$$= 1/(2)(3.14)(10^2 Hz)(0.707)(10^{-8} F)(2(1)+1)$$
$$= 1/(2)\ (3.14)\ (10^2)\ (0.707)\ (10^{-8})\ (3)$$
$$= 1/(1.332 \times 10^{-5}) = 75.076\ ohms$$

Calculate R2–

$$R2 = Q(2A_{vb} + 1)/2 \pi f_c C_1$$
$$= (0.707)(2(1) + 1)/(2)(3.14)(10^2)10^{-8} F)$$
$$= 2.121/6.28 \times 10^{-6} = 337,739\ ohms$$

Bandpass MFPF Filters

Figure 16-14 shows an example of the bandpass MFPF filter. This is the circuit mentioned at the end of the VCVS filter section that uses only one operational amplifier to generate a −12 dB/octave, second-order, bandpass filter response. This circuit

178

appears at first glance to have elements of the low-pass and high-pass MFPF designs. The bandpass gain of this circuit is given by:

$$A_{vb} = 1(R_1 R_3)(1 + C_2/C_1)$$

If you specify the maximally flat version of this circuit, set C1 = C2, so this equation reduces to:

$$A_{vb} = 1/(2R_1/R_3$$
$$= R_3/2R_1$$

The circuit Q, which is defined as $f_o/(f_h - f_L)$, determines the sharpness of the circuit and should be determined from the application in which the filter is to be used. The center frequency is found from

$$f_o = \frac{1}{2\pi C_1} \sqrt{\frac{(R_1 + R_2)}{R_1 R_2 R_3}}$$

The resistor values are calculated from the following expressions:

$$R1 = Q/(A_{vb}^2 \pi f_o C_2)$$
$$R2 = Q/(2Q^2 - A_{vb})(2\pi f_o C)$$
$$R3 = 2Q/\omega_o C_2$$
$$R3 = Q/\pi f_o C_2$$

Determine the correct values using the following sequence:

☐ Determine Q, f_o, f_L and f_h from consideration of the application.

☐ Set C1 to a convenient value.

☐ Set C2 = C1.

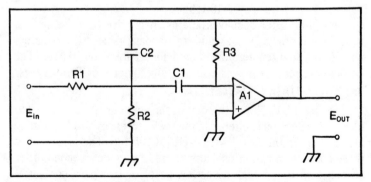

Fig. 16-14. Bandpass MFPF circuit.

☐ Compute R1.
☐ Compute R2.
☐ Compute R3.

☐ **Example:**

Design a 200-Hz filter with a Q of 12. Assume a bandpass gain of unity.

Determine f_L and f_h (already specified are $f_o = 200$ Hz and Q = 12) $Q = f_o/(f_h - f_L)$ so $(f_h - f_L) = f_o/Q = 200\,Hz/12 = 16.7\,Hz$. If you assume that the bandpass is symmetrical about f_o, then:

$$f_h = f_o + (16.7/2) = 200 + 8.4\,Hz \cong 208\,Hz$$
$$f_L = f_o = (16.7/2) = 200 - 8.4\,Hz \cong 192\,Hz$$

Set C1 to a convenient value. C1 = 0.05 μF.
Set C2 = C1. C2 = 0.05 μF.
Calculate R1—

$$R1 = Q/(A_{vb}\,2\,\pi\,f_o C_2)$$
$$= (12)/((1)(2)(3.14)(200)(5 \times 10^{-8}F)$$
$$= (12)/(1.256 \times 10^{-5}) = 79618\,ohms$$

Calculate R2—

$$R2 = Q/(2Q^2 - A_{vb})(2\,\pi\,f_o C_2)$$
$$= (12)/((2)(12)^2 - 1)((2)(3.14)(200)(5 \times 10^{-8}F)$$
$$= (12)/((286)(6.28 \times 10^{-5}))$$
$$= (12)/(1.796 \times 10^{-2}) = 668\,ohms$$

Calculate the value of R3—

$$R3 = Q/\pi\,f_o C_2$$
$$= (12)/(3.14)(200)(5 \times 10^{-8}F)$$
$$= (12)/(3.14 \times 10^{-5})$$
$$= 382,166\,ohms$$

As in all filter design problems, it is wise to make several trials in order to find values close to standard values. The resistors should be low-temperature coefficient, precision types. The capacitors should be polycarbonate, polyethylene, or Teflon types.

UNIVERSAL STATE VARIABLE FILTER

Figure 16-15 shows the circuit for a universal state variable filter. This version is offered in microelectronic hybrid IC form by Datel. This is the Datel Model FLT-U2 filter. The circuit uses three committed operational amplifiers to provide a second-order response in either high-pass, low-pass, or bandpass configurations. There is also one uncommitted operational amplifier that can

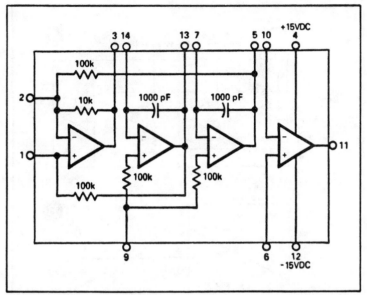

Fig. 16-15. State variable filter by Datel.

be gain programmed using the normal rules for operational amplifiers. Both the inverting and noninverting versions can be implemented.

One aspect to this filter type is that the low-pass, high-pass, and bandpass characteristics are available *simultaneously*. Several FLT-U2 filters can be cascaded to form-filter circuits with Q values

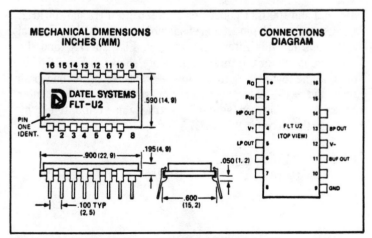

Fig. 16-16. Package and pinouts of the Datel state variable filter.

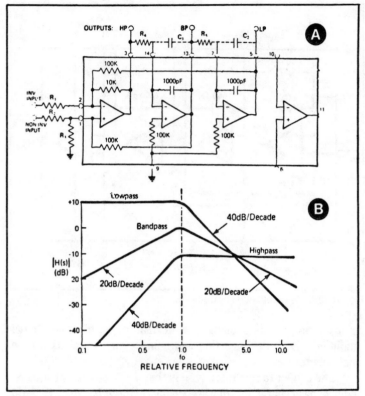

Fig. 16-17. FLT-U2 block diagram (A) and relative gains of simultaneous output, or Q = 1 (B).

between 1 and 1000 at frequencies between 0.001 Hz and 200 kHz. Package style and pinouts are shown in Fig. 16-16. Typical connections for this filter are shown in Fig. 16-17A, and the respective gains for the three outputs are shown in Fig. 16-17B. Note that the low-pass output gain is +10 dB above the bandpass output and +15 dB above the high-pass output at the center frequency, f_o. You can see this is true almost intuitively by examining the circuit in Fig. 16-17A.

Chapter 17
AC Amplifiers

In the simplest terms, an AC amplifier is one which is sensitive to AC signals, but not DC signals. This does not, in any way, mean that DC amplifiers will not handle AC signals. In fact, most DC amplifiers have a frequency response from DC to some specified frequency. An AC amplifier, on the other hand, will respond to only AC, which needs some qualification. The qualification on the definition of AC is that the amplifier will respond to varying signals that are unidirectional or unipolar. These might be considered varying DC potentials in some books, but they are processed in AC amplifiers much of the time. Many scientific instruments, especially those in the biophysical or medical fields, are designed to use AC amplifiers, even though the low end of the frequency response is on the order of less than 1 Hz. The electrocardiograph (ECG) machine, for example, uses an AC amplifier with a frequency of only 0.05 Hz as the low end limit. The reason for using a low frequency that is so near DC is that the waveform is a very low frequency signal, yet there are problems with DC offset that must be overcome. The ECG electrode is fastened to the patient's skin through a conductive gel. The skin-gel electrode interface forms a miniature battery that can produce potentials in the over-1 V range. These potentials are not desired, yet some of the frequency components of the desired signal are close to DC. As a result, AC amplifier has a frequency response down to 0.05 Hz.

There are other reasons why one would use an AC amplifier. We might, for example, want to limit either the low end or upper end frequency response in order to get rid of certain types of noise signal. This is done in certain instrumentation applications where a carrier amplifier is used to process the analog data. Care must be taken, however, to make sure that the frequency response limits are selected with some knowledge of the signal that will be processed. Remember that all periodic, continuous functions have component frequencies that are described by the Fourier series of that function. If the frequency response is limited too much, we will have a distortion of the signal that could be very bad. This is, incidentally, why various standards groups have set the frequency responses required of certain classes of instruments. To make everything consistent, the various medical/hospital authorities recognize a need for frequency response limits of 0.05 to 100 Hz for ECG amplifiers.

AC AMPLIFIER CIRCUITS

Figure 17-1 shows one of the easiest methods for building an AC amplifier. Note that all three circuits in Fig. 17-1 are based upon the simple DC amplifiers of earlier chapters, which is both a strength and a weakness.

The circuit shown in Fig. 17-1A uses a noninverting follower operational amplifier as the basis of the configuration. This circuit is converted from DC to AC operation by interstage transformer T1. The output voltage of the amplifier is the product of the gain of the operational amplifier and the step-up ratio of the interstage transformer. This transformer step-up ratio is merely the ratio of secondary to primary turns and must be accounted for in the transformer equation:

$$A_v = \left[\frac{N_s}{N_p}\right]\left[\frac{R_f}{R_{in}} + 1\right] \qquad \textbf{Equation 17-1}$$

where N_s is the number of turns in the transformer secondary, N_p is the number of turns in the transformer primary, R_f is the resistance of the feedback resistor, and R_{in} is the resistance of the input resistor.

□ **Example:**

Calculate the voltage gain of an amplifier such as the one shown in Fig. 17-1A if the transformer has a 1:3 turns ratio, the feedback resistor is 100K ohms and the input resistor is 10K ohms.

$$A_v = \left[\frac{N_s}{N_p}\right]\left[\frac{R_f}{R_{in}}\right] + 1$$

$$= \left[\frac{3}{1}\right]\left[\frac{100}{10}\right] + 1$$

$$= (3)(10+1) = (3)(11) = 33$$

There are two problems with this circuit. One is that transformers are imperfect devices and may have insufficient frequency response for some applications. The other problem is that transformers are usually made larger than other components in the circuit, and this may be difficult to accommodate in some applications.

Figure 17-1B shows an AC amplifier adapted from the simple DC inverting follower circuit. In this particular version of the circuit, capacitor C1 is used to block DC, yet it will allow AC signals to pass. There is, though, some attenuation of AC signals at frequencies for which the reactance is high. Although this may seem to be a disadvantage in some applications, it is a decided advantage in those cases where we want to tailor the low-frequency response.

Another version of this same concept uses the noninverting amplifier configuration. It is shown in Fig. 17-1C. Again, the purpose of the capacitor is to block DC while passing AC signals. In an ideal circuit using ideal components we would not need resistor R1 (remember that the input impedance of an operational amplifier is infinite). But real operational amplifiers are far from ideal: They often produce input bias currents. These currents, though small, will tend to charge input capacitor C1. In time (often only a few seconds), the charge on capacitor C1 will totally block the operational amplifier. To the operational amplifier, this voltage is another valid signal, so it will produce an output voltage that is proportional to the product of the gain of the amplifier and the voltage across capacitor C1. The operational amplifier will tend to

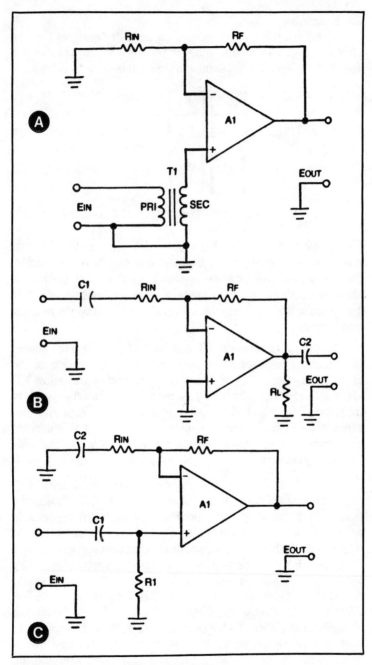

Fig. 17-1. Various forms of AC-coupled amplifier.

become latched up, even though the bias is very small. Resistor R1 is used to provide a discharge path for capacitor C1, although it is at the expense of reducing the amplifier input impedance. In the ordinary noninverting follower, the input impedance is very high (sometimes in the terraohm range), but it drops to the value of R1 in the circuit of Fig. 17-1C. We want to select a value for R1 that is very high with respect to the source impedance of the preceding stage, yet is low enough to permit the antilatchup operation which we described.

An improved AC amplifier that makes use of the feedback properties of an operational amplifier is shown in Fig. 17-2. In many applications, this is the circuit of choice.

An IC gain block called the *current difference amplifier* (CDA), or *Norton amplifier* (see Chapter 18) is also useful as an AC amplifier. The CDA is in fact almost always used in the AC configuration due to certain aspects of its basic design.

VERY LOW FREQUENCY RESPONSE OPERATION

The frequency response properties of most amplifiers is determined by the values of the resistors and capacitors in the circuit, or sometimes by the upper frequency limit of the amplifying element. There is a rule-of-thumb for selecting the values of most bypass, decoupling, or coupling capacitors. The rule requires a capacitor value that has a reactance at the lowest frequency of operation of one-tenth the associated circuit resistance. This requirement presents a bit of a problem when the low-end frequency limit is very low, such as in physiological

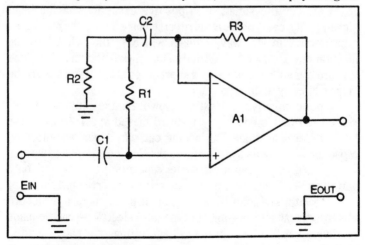

Fig. 17-2. Bootstrapped AC amplifier.

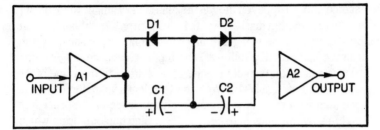

Fig. 17-3. Method for using moderate value capacitors to obtain low-frequency response.

amplifiers with a 0.05 Hz lower −3 dB point. Because capacitive reactance is inversely proportional to frequency, a large value capacitance is needed at those frequencies. This situation means that we will need an electrolytic (aluminum or tantulum) capacitor for such applications, and these are polarity sensitive. Such sensitivity is useless in an AC amplifier, of course.

A method used to overcome the problems of electrolytic capacitors in some medical equipment is shown in Fig. 17-3. The high capacitance is retained, yet the circuit is comfortable with bipolar AC signals. If you look at this circuit too quickly, you might conclude that it is merely a series connection of two electrolytics. If that were true, the total capacitance would be equal to C/2, provided that C1 = C2. In reality, though, the total capacitance in the circuit of Fig. 17-3 is C, or the value of either C1 or C2. Again, the assumption is that equal value capacitors are used. This action occurs because of D1 and D2.

Consider the situation on positive excursions of the input signal. In this case, diode D1 is reverse biased, so it does not affect capacitor C1 in any way. Signals will pass through C1 to the junction of C1-C2=D1-D2. Diode D2, on the other hand, will be forward biased, so it effectively shorts capacitor C2 and passes the signals directly to the next stage.

On the negative alternation, however, the situation is reversed, so that diode D1 is forward biased and D2 is reverse biased. Because diode D2 is now cut off, it has no effect on capacitor C2. Signals will pass through C2 to the following stages.

From this, you can see that each capacitor is in the circuit only during one-half of the cycle. The polarities are arranged such that each capacitor sees only the signals of the proper polarity. Modern electrolytic capacitors—in both types of dielectrics—can be made relatively small in size in the voltage ratings normally found in operational amplifier circuits.

Chapter 18
CDA and OTA Amplifiers

The operational amplifier probably revolutionized analog electronics design and is still the mainstay of many electronics designers. But there are several other types of linear IC amplifiers. Some of them are merely various types of transistor arrays, and these will be discussed in the Chapter 19. Others are specialized devices that obey operational amplifier-like rules but are definitely not classical op amps. Two of the most popular of these devices are the *current difference amplifier* (CDA, and also called the Norton amplifier) and the *operational transconductance amplifier* (OTA). These are discussed in this chapter.

CDA DEVICES

The classical operational amplifier that was discussed in previous chapters is a voltage difference device. There are two input terminals—inverting and noninverting which have equal but opposite effects on the output signal; that is, they produce equal amplitude signals that are 180 degrees out of phase when the same input signals are applied.

The current difference amplifier, or CDA, is similar to the operational amplifier in many respects, but it is a current-input device. These amplifiers are operated from a monopolar DC power supply, so they are suited well to the mobile electronics equipment market.

The symbol for a CDA is shown in Fig. 18-1. Note that it is the same as the classical operational amplifier symbol with a constant current source (CCS) symbol along the back edge of the triangle.

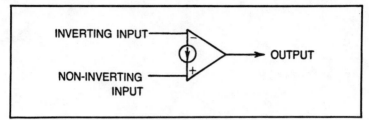

Fig. 18-1. Symbol for a current difference amplifier (CDA).

The CDA can be discussed in its own right. However, it is sometimes instructive to consider the CDA in light of comparison with the classical operational amplifier.

A simplified schematic of a voltage-difference op amp input stage is shown in Fig. 18-2A; the op amp input stage is shown in Fig. 18-2B. This circuit is a differential amplifier consisting of a constant current source (CCS), usually a transistor in its own right,

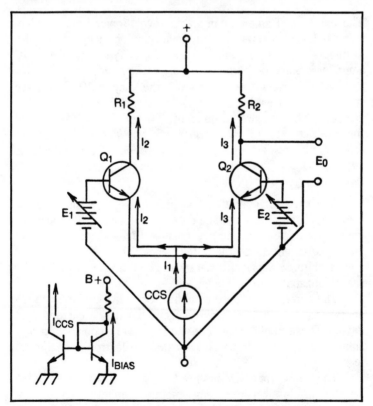

Fig. 18-2. Typical CDA input stage compared with op amp input stage.

and a differential pair of transistors Q1 and Q2. These transistors are a matched pair, and their emitters are connected together. Current I1 is constant, so when the input voltage to the transistors are equal, currents I2 and I3 will be equal to each other. In that particular case, because I1 = I2 + I3, and I2 = I3.

If either I2 or I3 is changed, the other will also change, in order to maintain the constancy of I1. For example, if E1 is increased, current I2 will also increase. This will cause current I3, however, to decrease an amount equal to the change of I2. Voltage output V_o is taken from the collector of Q2. When E1 increases, causing a decrease in I3, the voltage at the output will also increase. Therefore, the base of Q1 is used as a *noninverting input*.

The base of transistor Q2 is the inverting input. If voltage E2 is increased, the collector current of Q2 is increased, and this will cause V_o to decrease. This is due to the same mechanism discussed in the early chapters: The voltage drop across the collector load resistor increases and decreases with collector current variations.

Figure 18-3 shows a simplified version of a typical CDA input stage. First consider the operation of the inverting input. Transistor Q1 is a common-emitter npn amplifier with a constant current source in the collector circuit. This transistor feeds the base of transistor Q2, which is used as an output buffer. The emitter circuit of Q2 also contains a constant current source.

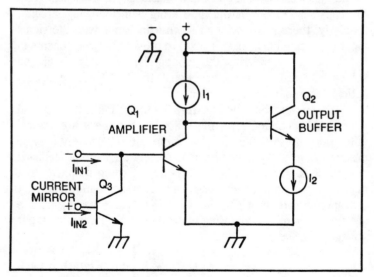

Fig. 18-3. Typical CDA input stage expanded.

The noninverting input is formed through the use of a *current mirror* (transistor Q3). A current input to the noninverting terminal—the base of Q3—is subtracted from the current applied to the inverting input through the sinking action of the Q3 collector. Hence, the current seen by the input transistor Q1 is the difference between the inverting and noninverting input currents. In most cases, the noninverting input is given a permanent bias of 5 to 100 μA. This bias is provided usually by a resistor to the V+ power supply.

A more detailed view of the input amplifier is shown in Fig. 18-4. This particular circuit shows the constant current source transistors and the input diode. Bias to the CCS transistors is provided by an internal regulator circuit. It is possible to achieve quite decent voltage regulation internally in ICs and gain the added advantage that the changing thermal environment affects both the amplifier and regulator components equally. The current remains stable because the internal bias conditions are stable. In both cases, the collector current of the CCS transistors are used as the collector currents for the active transistors. Note that negative polarity signals look into a diode at the + input terminal.

INVERTING AMPLIFIERS

The CDA is inherently an AC amplifier because there is a constant DC level present at the output terminal. In most cases, we would want a capacitor at the output of the CDA in order to prevent the constant DC level from interfering with following circuits. Similarly, there is a blocking capacitor in series with the signal input (inverting input, in this case) to prevent DC components in the signal from affecting the input. The capacitor should be selected to have a reactance equal to $R_{in}/10$ at the lowest anticipated frequency of operation.

This circuit, shown in Fig. 18-5, looks much like the classical operational amplifier inverting follower, but there are critical differences. In the operational amplifier, it is relatively easy to predict the gain of the circuit; it is merely the ratio of feedback resistor R_f to input resistor R_{in}. It is this startling simplicity that makes the op amp so appealing to many. This same ratio yields only the approximate gain for the CDA; several error terms are accounted for by the following equation for the quiescent output voltage:

$$E^1_o = \frac{E_{ref}R_f}{R_{ref}} + \left[1 - \frac{R_f}{R_{ref}} \right] \phi \qquad \textbf{Equation 18-1}$$

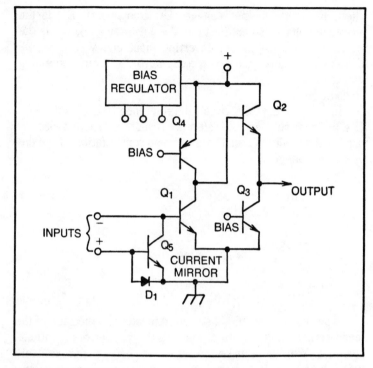

Fig. 18-4. Typical CDA.

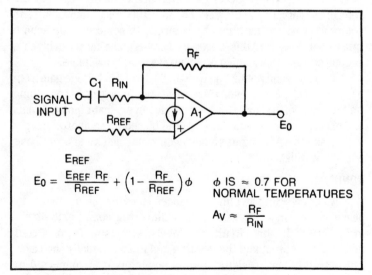

$$E_0 = \frac{E_{REF}\ R_F}{R_{REF}} + \left(1 - \frac{R_F}{R_{REF}}\right)\phi$$

ϕ IS ≈ 0.7 FOR NORMAL TEMPERATURES

$$A_V \approx \frac{R_F}{R_{IN}}$$

Fig. 18-5. Typical circuit using a CDA device (inverting).

where E^1_o is the output voltage (DC component), E_{ref} is the reference voltage (sometimes the V+ power supply), R_{ref} is the resistance between the noninverting input and V_{ref}, R_f is the feedback resistance, and ϕ is a constant (0.7 at normal operating temperature).

☐ **Example:**

In an inverting CDA circuit the reference voltage is the +15 VDC power supply, and the reference resistor is exactly twice the value of the feedback resistor (a common configuration). Find the DC output voltage.

$$E^1_o = \frac{E_{ref} R_f}{R_{ref}} + (1 - R_f/R_{ref}) \; \phi$$

$$= \frac{(15V)(R_f)}{(2R_f)} + (1 - R_f/2R_f)(0.7)$$

$$= (15V)(\tfrac{1}{2}) + (1 - \tfrac{1}{2})(0.7) = (7.5) + (0.5)(0.7) = 7.85 \text{ volts}$$

The AC voltage gain is set approximately by the ratio of the feedback and input resistors, much in the manner of operational amplifier. We like to make the reference resistor approximately twice the feedback resistor but are constrained somewhat by a desire to keep the value of the reference resistor such that the reference current is between 5 and 100 microamperes. The reference resistor may also be returned to some voltage source other than V+, but this becomes cumbersome as it involves a second power supply which is not always readily available.

We can usually obtain up to 60 to 80 dB of AC voltage gain from a typical CDA device, even though the open-loop gain is something on the order of 100 dB. Considered, however, that bipolar signals are handled only because the input is biased up to approximately ½V+, and the AC output signal swings plus and minus, or above and below, this level.

NONINVERTING AMPLIFIERS

A noninverting amplifier produces an output signal that is in phase with the input signal. A CDA noninverting amplifier is shown in Fig. 18-6. In this circuit, the feedback resistor is connected between the output and the negative input, as is usually the case. But note the input resistor. In this amplifier, it is connected in series with the noninverting input. As in the previous case, the reference resistor is connected between the V+ power supply (or a reference power supply) and the noninverting input. This amplifier inputs the signal through the current mirror. See Fig. 18-7. The gain of the circuit for AC signals is given by:

$$A_v = R_f/(R_{in}(26/I_{ref}))$$

where R_f is the feedback resistor, R_{in} is the input resistor, and I_{ref} is the reference current in milliamperes.

DESIGN PROCEDURE FOR CDA CIRCUITS

For most designs, the first step will be selecting the quiescent bias current. This should be some value between 5 μA and 100 μA. The bias current is selected. Then a resistor must be calculated that will deliver this current. The resistor is usually tied from the noninverting input of the CDA to either a reference supply or the V+ power supply, and it will have a value of V+/I_{bias}.

In most cases, it is convenient to set the quiescent output voltage to ½V+. For this particular case, the feedback resistor should have a value that is equal to one-half the value of the bias resistor selected. Some amount of dickering can be done between the quiescent current and the resistor values so that a 2:1 match is possible.

The selection of overall gain is a matter to consider before beginning to design the circuit. It will most likely be dictated, at least in general terms, by the needs of the particular application intended. Once you have the desired gain and know the value of the feedback resistor, you can select an input resistor from the equations given earlier.

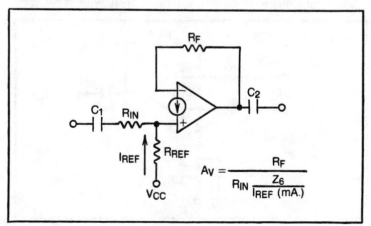

Fig. 18-6. Noninverting CDA circuit.

195

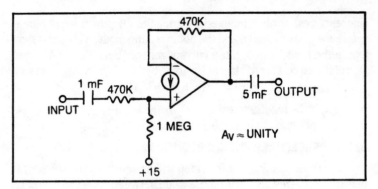

Fig. 18-7. Another noninverting CDA circuit.

DIFFERENTIAL CDA AMPLIFIER CIRCUITS

The differential amplifier was first considered in the chapters on operational amplifiers. In those devices, the output voltage was proportional to the difference between two input voltages, one each applied to the inverting and noninverting input terminals. The differential amplifier finds wide application in instrumentation, medical/scientific electronics and many other applications. Other applications are for the differential current input amplifier, or CDA. In these circuits, the output is proportional to the difference between the currents flowing into the inverting and noninverting inputs of the CDA device.

Figure 18-8 shows how the current-input CDA can be used for making a differential voltage amplifier. Take advantage of Ohm's law and place a pair of resistors in series with the two CDA inputs. The output voltage will be proportional to the difference between

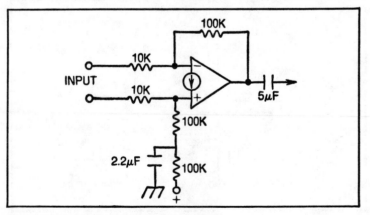

Fig. 18-8. Differential amplifier using CDA circuit.

the currents flowing in the two input resistors, which are in turn proportional to the applied differential voltage. Note that an output DC blocking capacitor is used to prevent the ½V+ DC component on the output terminal from interfering with the operation of the stage that follows.

AUDIO MIXERS

An audio mixer is a circuit that combines the signals from two or more sources in a linear manner. Audio and radio broadcasting engineers use mixers to combine several program sources, including microphones, phonographs, tape recorders and air pickups from radio receivers and monitors. The CDA is a seemingly universal AC amplifier, so it is almost ideally suited to such applications. Figures 18-9 and 18-10 shows two different approaches to the design of CDA audio mixer amplifiers.

The more complex, and presumedly more preferred, circuit is shown in Fig. 18-9. In this circuit, a master amplifier (one CDA) is driven by the outputs of several input preamplifier stages (other CDAs). A quad-CDA IC device, such as the LM3900, works nicely as a three-channel mixer using this system. Note that only one preamplifier channel is shown in this circuit, but the other stages will be identical to A1 in Fig. 18-9.

One appealing feature of this circuit is that it permits each channel to be turned on and off with a simple switch that handles the DC bias current for the stage. When the noninverting input is grounded, the preamplifier CDA is turned off because the quiescent bias is zero. When the switch is open, however, the bias current is allowed to flow, so the amplifier operates. It is generally considered best to switch DC control voltages or currents instead of AC signals, because of the many problems caused by running AC signal lines all over the equipment.

The mixer circuit shown in Fig. 18-10 is appealing for its utter simplicity. Note that this circuit is much like similar circuits made from ordinary voltage operational amplifiers. The CDA is biased in the normal manner, and a feedback resistor is provided that has a value of approximately one-half of the bias resistor value. Note that this is not exact; standard values have been selected for convenience. This makes the circuit easier to build and results in only a smaller error in the quiescent output voltage.

The input resistors are selected in the ordinary manner, using the formula given earlier. In most cases, select the same value for all of the resistors, unless there is a need to interface signal

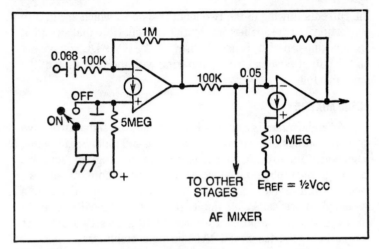

Fig. 18-9. Audio mixer circuit using CDA devices.

sources of vastly different amplitude levels. In that case, select values that would accommodate the various signal sources.

All of the input signals are combined in a summing junction at the input side of the input coupling capacitor. The output voltage will be the summation of the input contributions, considering their respective gains. Individual level controls in Fig. 18-10 can be provided by using potentiometers either for the input resistance or to drive the input resistor. If you desire individual control over levels, however, it might be wise to use the circuit of Fig. 18-9 and provide some isolation between the circuits.

OTHER CDA CIRCUIT APPLICATIONS

In this section, we will consider some of the miscellaneous circuit designs that might prove useful. These are not treated in a

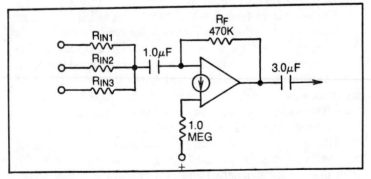

Fig. 18-10. Another Audio mixer circuit using CDA devices.

198

comprehensive manner, but are merely representative of the possibilities of the CDA. You will no doubt also see some other possibilities which apply to your own needs.

Figure 18-11 shows a low-level DC preamplifier circuit. DC signals are notoriously difficult to handle when their amplitude, or value, becomes similar to the offset voltages and/or currents in the amplifying devices. Low-level signals also have difficulty biasing a transistor into operation. This is especially common on circuits such as the common-emitter amplifiers used as the input stages of the CDA. This problem is overcome in Fig. 18-11 with bootstrapping resistors connected to the V+ power source, and the resistor to ground from the noninverting input terminal.

A simple voltage comparator is shown in Fig. 18-12. A *comparator* is a circuit that provides a means for comparing two voltages and then issues an output that tells whether they are equal, or which of the two is greater. Most comparators are merely amplifiers with excessive gain. In that case, a differential voltage of only a few millivolts (or even *micro*volts) will cause the output to saturate at either the V+ or V− supply rails. As in the case of operational amplifier comparators, the CDA circuit of Fig. 18-12 achieves the excessive gain by eliminating the feedback resistor. The gain of this circuit is the open-loop gain of the particular CDA device selected. Assuming that the input and bootstrap resistor equalities are maintained (R1=R2 and R3=R4), the output will be zero when the unknown voltage, V_x, is equal to the known reference voltage, V_{ref}. Under this condition, the two input currents are equal, so the output voltage is zero. A difference between the two input voltages, however, will generate differential input current, so the output voltage must be nonzero. The excessive gain of the circuit makes the output voltage equal to the maximum allowable voltage for the V+ applied. When only a single monopolar power supply is used for the comparator, only one quadrant of operation is allowed. If two-quadrant operation is desired, which would tell us not only that the voltage were unequal, but which was greater, the circuit would have to operate from a bipolar power supply as op amps do.

Schmitt triggers are circuits that will square an input waveform (Fig. 18-13). The output remains low until the input signal voltage exceeds a certain preset threshold. It then snaps high and remains high until the input voltage falls below a second threshold. The first and second threshold voltages are rarely equal. An example of a CDA Schmitt trigger is shown in Fig. 18-14. The input signal is

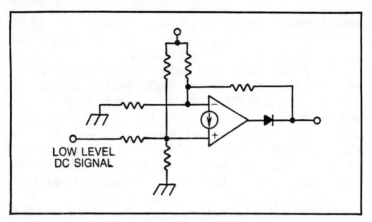

Fig. 18-11. Low-level DC amplifier based on the CDA.

applied to the inverting input through a 1 megohm resistor. The feedback resistor is connected to the noninverting input, resulting in a little positive feedback. For reasons that we will not discuss here, this is the cause of the Schmitt action.

OPERATIONAL TRANSCONDUCTANCE AMPLIFIERS

The operational transconductance amplifier is an op amp-like device originated by RCA several years ago. The classic operational amplifier is a voltage amplifier. It has a transfer function that relates an output voltage to an input voltage. The CDA was a current amplifier; it related an output current to an input current. The CDA could also be a transresistance amplifier because it related an output voltage to an input current. The definition of a transconductance amplifier is exactly the opposite of the resistance amplifier: It has a transfer function that relates an output current to

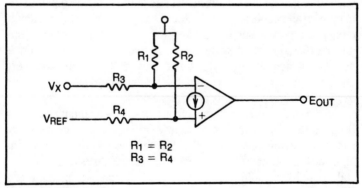

Fig. 18-12. CDA circuit.

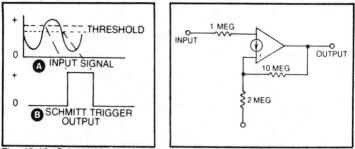

Fig. 18-13. Schmitt trigger characteristic. Fig. 18-14. Schmitt trigger circuit.

an input voltage. The units for measuring transconductance are naturally enough 1/R units, called *mhos*. We calculate transconductance from:

$$g_m = \frac{\Delta I_o}{\Delta E_{in}}$$

where g_m is the transconductance in mhos, I_o is the output current, E_{in} is the input voltage, and Δ denotes a small change in the associated parameter.

☐ **Example:**

An amplifier exhibits a 1mA change in output current I_o cuased by a change of input voltage of 0.2 volts. What is the transconductance?

g_m = $\Delta I_o / \Delta E_{in}$

= $(0.001\,A)/(0.200\,volts)$

= 0.005 mhos x 10^6 μmhos/ mho = $5000\,\mu$mhos

The circuit symbol for the OTA is the same as for the ordinary operational amplifier, although it is probably advisable to use the letters "OTA" inside of the triangle to let it be known that it is not an op amp device. One advantage of the transconductance amplifier over the standard voltage operational amplifier is that the transconductance (gain) of the amplifier can be set by the designer through the selection of a current for a bias terminal (see the RCA specifications sheet for the CA3080 device). This one feature adds a certain amount of flexibility to the OTA.

Chapter 19
Linear IC Amplifiers

Integrated circuitry (IC) lends itself to the construction of all manner of amplifier circuitry. In fact, nonoperational amplifiers were among the first uses of the new technology called integrated circuitry—way back in the early 1960s. The litte μA703 device, for example, preceded the first IC operational amplifier (μA709) and was bipolar amplifier that came to be used in many FM radio i-f amplifier stages. Over the years, a number of different type of linear ICs have been developed that are neither operational amplifiers nor CDA/OTA devices.

Also, many devices a called *hybrid amplifiers*. The hybrid is a cross between integrated circuitry and discrete circuitry. An example of a hybrid circuit module is shown in Fig. 19-1. The substrate is a ceramic-base printed circuit, and unpackaged integrated circuit chips are mounted on the ceramic. Capacitors are formed either from the PC material, or by mounting chip capacitors onto the substrate similar to the way the IC chips are mounted. Very fine gold wires are used to connect the pins of the package to the various points on the integrated circuit. Although not shown in Fig. 19-1, the final product will have a hermetically sealed case or will be completely potted in epoxy resin. The hybrid function module might be an operational amplifier, or a linear amplifier for some specific purpose. At least one product made by Burr-Brown is an amplifier that has a transfer function of the form $X(Y/Z)^m$, where X, Y and Z are voltage inputs and m is a value set by a

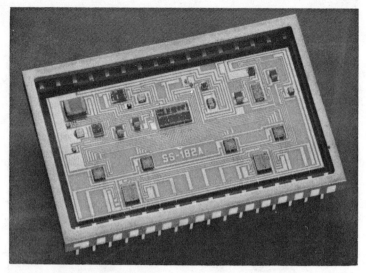

Fig. 19-1. Hybrid circuit module.

resistor network from 0.2 to 5. If the resistor network is 0.5, the transfer function will take the square root of the input voltage Y/Z. Similarly, powers up to $(Y/Z)^5$ can be obtained. If we want to make the device behave like an ordinary linear amplifier, we would select a resistor network with a value of 1, and then set the Y and Z inputs to the same constant voltage so that $Y/Z = 1$.

In this chapter we are going to consider a few of the commercial products that are available on the market. The available space does not allow us to consider all of the known products, but we will cover a representative sample. Also, not all applications of any given product can be covered. Some devices are used for a wide range of applications; some of the applications are not even on the list dreamed up by the original product developers!

DIFFERENTIAL AMPLIFIER CIRCUITS

The basic transistor differential amplifier forms the basis for most linear IC amplifiers, including operational amplifiers. Figure 19-2 shows the basic circuit for a differential amplifier. Transistors Q1 and Q2 form a *differential pair*, meaning that they are not only identical as to type number but have also been matched as to specific parameters. In an integrated circuit, a differential pair works very well because they are formed onto the same substrate of the same batch of material at the same time. They are inherently closely matched. Also, they share the same thermal environment, so they will remain matched when the temperature changes. This is an immense advantage of the IC differential amplifier over the discrete types.

Notice that the emitters of transistors Q1 and Q2 are connected together, and are fed from a *constant current source* (CCS). The constant current source has the effect of keeping current I3 at the same value (constant), regardless of changes in load impedance occasioned by changes in Q1 and Q2.

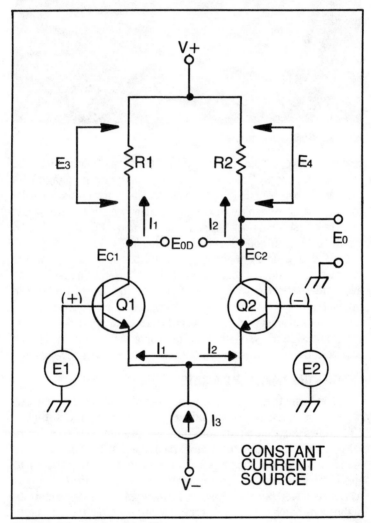

Fig. 19-2. Differential input amplifier.

204

At this point, let us maintain a little fiction regarding the respective collector and emitter currents of Q1 and Q2. We will assume that the collector and emitter currents for Q1 (I1) and Q2 (I2) are equal. Similarly, the current in the collector of Q2 is equal to the current in the emitter of Q2. This is near enough to the truth as to not offend reality. Besides, the collector-emitter current ratios *are* equal!

Now that we have set up a ground rule that will facilitate our thumbnail analysis, let us proceed with a description of the circuit operation. We know from Kirchoff's current law (KCL) that constant current I3 is equal to the sum of the two emitter currents:

KCL: $$I3 - I1 - I2 = 0$$

Equation 19-1

Therefore: $$I3 = I1 + I2$$

Under conditions where base voltages E1 and E2 are equal, the two emitter currents are equal. This fact follows from the properties of bipolar transistors:

$$I1 = I2$$ **Equation 19-2**

The voltage at either of the two collectors is less than the supply voltage and will depend upon the value of the collector current. For transistor Q1, collector voltage E_{C1} is the supply voltage less the voltage drop across collector load resistor R1:

$$(V+) - (I_1 R1) = E_{C1}$$ **Equation 19-3**

Similarly, the collector voltage of transistor Q2 is the difference between the supply voltage V+ and the voltage drop across resistor R2:

$$(V+) - (I_2 R2) = E_{C2}$$ **Equation 19-4**

If we set R1 = R2 = R, these expressions can be rewritten:

$$E_{C1} = (V+) - (I_1 R)$$ **Equation 19-5**

and

$$E_{C2} = (V+) - (I_2 R)$$ **Equation 19-6**

The single-ended output voltage, as shown in Fig. 19-2, is the collector voltage of transistor Q2; in other words, output voltage

E_o is E_{C2}. The differential output voltage is taken between the two collector terminals. In this case, the differential output voltage is $E_{C2} - E_{C1} = E_{od}$.

Let us examine the differential output voltage under the condition where $E1 = E2$. We know that, under that condition, $I1 = I2$, so let's set $I1 = I2 = I$ and substitute into some equations:

$$E_{C1} = E_{C2}$$ **Equation 19-7**

$$(V+) - (IR) = (V+) - (IR)$$ **Equation 19-8**

$$0 = 0$$ **Equation 19-9**

Clearly, then, the output voltage for the differential case (E_{od} in Fig. 19-2) is zero when the two input voltages are equal. This justifies the use of the terminology, "differential amplifier," for this circuit. The definition of a *differential amplifier* is that it produces an output only when the difference between the two input voltages is non zero.

Now, examine the single-ended output voltage of the same circuit. Again, consider the case where $E1 = E2$. Again $I1 = I2$. The output voltage, however, is E_0 (see Fig. 19-2). This voltage is the same as E_{C2}, so use Equation 19-6 to calculate the output voltage:

$$E_o = (V+) - (I_2R)$$ **Equation 19-10**

Clearly, then, the output voltage under this circumstance is nonzero when the two input voltages are equal, *unless* the magnitude of I2 is sufficient to completely drop the supply voltage across the collector load resistor. This circuit behaves a bit more like ordinary transistor amplifiers.

The single-ended output configuration produces an interesting side effect. There are now two different types of input: inverting and noninverting. As in the case of operational amplifiers (most of which use a bipolar transistor or FET differential amplifier circuit for the input stage), the inverting input is labeled with a minus sign (−) and the noninverting input is labeled with a plus sign (+). But which is which? You can tell by answering the question, "Which input causes the output signal to be in phase, and which causes it to be out of phase, with the input signal?" Assume the case where E1 is greater than E2. This is the same as applying a positive voltage to that input. In that case, current I1 will increase.

But, since the KCL relationship of Equation 19-1 must be maintained, increasing I1 can only decrease I2: I3 must remain constant. When current I2 decreases, the voltage drop across R2 also decreases. Because the output voltage is $(V+) - (I_2 R)$, E_o will increase. We have now identified the noninverting (+) input. The base of transistor Q1 forms the noninverting input because a positive voltage applied to this point causes a positive-going output change at the collector of Q2.

Will we find next that the base of Q2 is the inverting (−) input? Apply input potentials such that E2 is greater than E1. This is equivalent to applying a positive voltage to the base of Q2. This situation causes current I2 to increase, thereby increasing the voltage drop across resistor R2. Output voltage E_o, therefore, decreases for an increasing input voltage. Hence, the base of Q2 is the inverting input.

RCA CA-3000 SERIES DEVICES

Some of the earliest and longest lived IC products that involve a differential amplifier are the devices sold under the RCA CA-3xxx series of type numbers. Some of these early integrated circuits are still available and are still considered a good choice for some applications. One of the simplest of these devices is a differential amplifier called the CA-3028A.

Figure 19-3 shows the internal circuit for the RCA CA3028A device. It is an unadorned differential amplifier using three transistors. Transistors Q1 and Q2 form the differential pair, while transistor Q3 can be used as a constant current source (if true diff-amp operation is desired) or can be used to modulate the output should multiplier, modulator or frequency mixer operation be desired. The base of transistor Q3 can be biased with the internal resistors (R1 and R2) if a potential is applied to pin No. 7 of the IC. Alternatively, the direct input (pin No. 2) can bias Q3. In many applications, input No. 7 will be used to establish a quiescent operating point. Then, a signal is applied to input No. 2.

The CA3028 can be used from DC to 120 MHz, and will dissipate up to 750 milliwatts if the temperature is less than 55°C, or 450 mW if the temperature is less than 85°C. The frequency capability of the CA3028 makes it useful for a wide range of applications. For example, the device can be used as a simple DC differential amplifier with a frequency response limited to a low value with external capacitors. In fact, if we want to use it only to some frequency less than a few kilohertz, the frequency limiting is all but mandatory to keep the thing from oscillating! The device can also be used as an rf or i-f amplifier at frequencies well into the low end of the VHF spectrum. In many amateur radio projects, the CA3028 is used as either an i-f or rf stage. The CA3028 comes in an eight-pin metal IC package, so it is relatively easy to use in practical applications.

An example of the type of circuit one would use for rf and i-f amplifiers is shown in Fig. 19-4. This circuit could be used in either rf or i-f cases, depending upon whether or not the tuned circuits are variable. The input signal is applied to integrated circuit U1 (an RCA CA3028) differentially across the bases of the two transistors (pins 1 and 5) of the differential pair. The input circuit is a resonant tank circuit with a link coupled transformer input. The tuned inductor of the input circuit L1 floats across the two IC inputs, as does the resonating capacitor. One side of the tank circuit (that

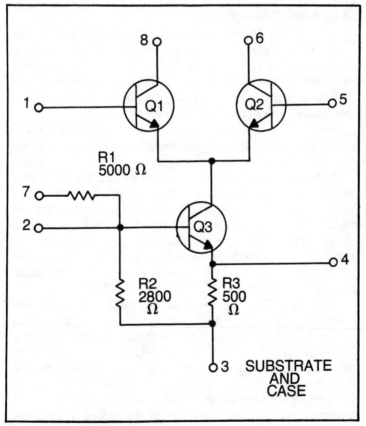

Fig. 19-3. RCA CA3028 device.

208

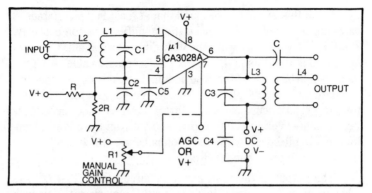

Fig. 19-4. CA3028 rf/i-f circuit.

connected to IC pin No. 5) is kept at AC ground by capacitor C2. This capacitor forms a bypass capacitor that keeps the IC input at AC ground, but above ground for DC. The DC bias is applied to this same point through an R-2R voltage divider circuit. In most cases, the value of R will be from 1K to 5K ohms, so the 2R resistor will be 2K to 10K ohms. Most frequently, 1K ohm and 2K ohm resistors are used in this application. The emitter of the internal current source transistor (Q3) is kept at AC ground by capacitor C5 connected to pin No. 4 of the IC. This capacitor is selected to have a capacitive reactance of 50 ohms (one-tenth the resistance of the emitter resistor R3) at the frequency of operation. In the case of a variable tuned amplifier, the reactance should be one-tenth the value of R3 at the lowest frequency of operation.

□ **Example:**

Select a bypass capacitor for pin No. 4 of an RCA CA3028 device that is intended for use as an rf amplifier in an 80-meter (3500-4000 kHz) amateur radio receiver.

We want the reactance to be 50 ohms, or less, at 3500 kHz. We will therefore use the capacitive reactance formula solved for C with $X_c = 50$ ohms.

$$C = \frac{1}{2\pi f X_c}$$

$$= \frac{1}{(6.28)(3.5 \times 10^6 \text{ Hz})(50 \text{ ohms})}$$

$$= 1/1.099 \times 10^9 \text{ farads}$$

$$= 9.1 \times 10^{-10} \text{ farads}$$

$$= 909 \text{ picofarads}$$

From this example, we found that at least 909 picrofarads of capacitance is required to make the amplifier function properly. But this is a value that would be very difficult to obtain. Our rule, however, is that this is the *minimum* acceptable value. A higher value would be even better. The next higher value capacitance in a standard table is 1000 pF, or 0.001 μF. We would select this value for a practical circuit. Note also that a 0.001 μF disc capacitor is very small—even in 1000V rating—so we gain the added advantage of circuit layout with small components, which can be very important in high-frequency circuits.

The output of the rf/i-f amplifier circuit (Fig. 19-4 still) is taken from pin No. 6 and is loaded by an LC tank circuit. This tank is tuned to the same frequency as the input tank circuit, so some caution in the layout of the circuit is in order. An alternate output is the capacitor shown, or we could use the popular link-coupled transformer method shown.

The internal resistor network that biases the current source (Q3) is connected to the outside world via pin No. 7 of the IC package. In this type of circuit, we may wish to connect the bias network directly to the V+ supply. The amplifier would then run wide open at maximum gain. Alternatively, we would gain a form of manual gain control by connecting pin No. 7 to a variable voltage source, such as a potentiometer (R1 in Fig. 19-4). Some radio receivers use a circuit called *automatic gain control* (agc), which keeps the output level of the radio relatively constant. These circuits sample the signal in the i-f amplifier and create a DC voltage that is proportional to the signal strength. This DC voltage is then used to control the gain of the rf amplifier—and sometimes the i-f amplifier, as well. Pin No. 7 of the CA3028A can be used as the agc terminal if the dc voltage from the agc circuit is applied as bias to transistor Q3. In short, the gain of the amplifier is controlled by controlling the conduction of transistor Q3.

The circuit shown in Fig. 19-4 uses the CA3028A in a relatively straightforward manner; i.e., the input is applied to the noninverting input and the output is taken in the single-ended manner. The circuit shown in Fig. 19-5A, however, is a little different. Let's examine first the simplified equivalent circuit in Fig. 19-5B. Note that the input stage is the current source transistor Q3, which is operated in the common-emitter mode. The output stage of this two-stage cascade amplifier is transistor Q2, which is operated in the grounded-base mode.

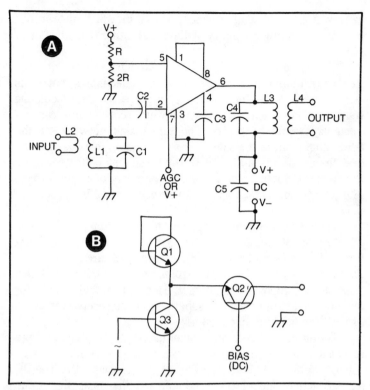

Fig. 19-5. Alternate rf/i-f circuit.

Now back to Fig. 19-5A, the actual circuit. The output stage is biased with the same type of R-2R resistor voltage divider as was used in the previous case. The input stage is biased also in the same manner as before. We could connect pin No. 7 either to V+ (wide open with maximum gain), to a manual gain control, or agc circuit. Once again, the emitter of transistor Q3 (pin No. 4) is bypassed to ground for AC. The same rule still obtains: The reactance of capacitor C3 should be 50 ohms or less at the lowest frequency of operation. Transistor Q1 is not used in this application, so pins 1 and 8 (the base and collector of Q1 respectively) are shorted together.

Figure 19-6 shows the CA3028A connected for use as a *multiplier, amplitude modulator* or *frequency mixer*. The tuned circuit makes it most useful as a mixer, but more of that in a moment. Be aware that all three of these applications are merely specific versions of each other. While the use as a multiplier might not be readily apparent, be well aware that modulation (AM) is merely a heterodyning mixing process in which the second frequency is a low-frequency audio signal rather than an rf or i-f signal.

The circuit of Fig. 19-6 has two inputs, one for each frequency. In this case, we are using two tuned inputs, but if this stage were a modulator, the input that receives the modulating signal would not be tuned, while that which received the "carrier" frequency would be tuned. The tuned tank network for the differential input is the same circuit configuration as used in the first circuit, shown in Fig. 19-4. The tuned tank is floated across the differential inputs and forms the secondary of a transformer. The actual input signal is applied to the primary of this transformer (L1).

The second signal is applied to the primary of a tank transformer that is capacitor coupled to pin No. 2. This configuration applies the signal developed across the tank circuit directly to the base of transistor Q3. By controlling current I3 (Fig. 19-3) this transistor can modulate the output. This input would be used for the local oscillator signal in a superheterodyne receiver or as the carrier in an amplitude modulator.

The output of this circuit is taken differentially by connecting a resonant tank circuit across the two transistor collectors. One collector (pin No. 8) is set to AC ground by capacitor C5. The DC for the "active half" of the differential amplifier is applied to the transistor through the tank circuit coil (L3).

VIDEO AND OTHER WIDEBAND AMPLIFIERS

Ever since television became a fact of life, we have been using video amplifiers. The *video amplifier* is used in television to handle

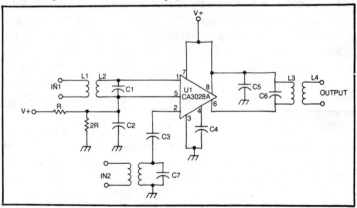

Fig 19-6. CA3028 Mixer amplifier.

212

the picture information either in the camera or in the receiver. When the vestigial sideband (a form of amplitude modulation) TV picture carrier is demodulated in the video detector of a television receiver, the filtered output will be video. This signal requires a wideband amplifier that is capable of passing a frequency spectrum to several megahertz. We generally accept 2.5 MHz for black-and-white monitors of the ordinary sort, 4.5 MHz for color receivers, and out to 15 MHz for certain types of high-resolution TV monitors used in security, instrumentation and medical X-ray fluoroscopy applications. But the video amplifier is little more than a wideband amplifier. The terms, wideband and video amplifier, can be used interchangeably if the particular wideband amplifier has the proper bandwidth in its frequency response.

Figure 19-7 shows a circuit for a wideband video amplifier based on the Motorola MC1590 integrated circuit, which is normally billed as an *analog multiplier*. One advantage of using integrated or hybrid circuit devices is the simpler external circuitry requirements. This device requires only a few external resistors and capacitors. Most transistor or vacuum-tube video amplifiers require rather substantial amounts of external components, including peaking coils and transformers.

Note the three points where capacitors are used. In all three cases, we have a parallel combination of a 1 μF and 0.001 μF capacitors. The point of this configuration is not to obtain a specific value (indeed, the tolerances on the 1 μF unit might be -20 to $+100$ percent!), but to accommodate both low-frequency and high-frequency signals. The 1 μF unit is an electrolytic (tantalum preferred), and these capacitors are not effective at high frequencies. At frequencies much above the audio range, the ability of the electrolytic capacitor to act as a capacitor is limited. Even in that range, a substantial series inductance is in some electrolytic capacitors, and this inductance will cause the performance of the capacitor to deteriorate. The low-value capacitor (0.001 μF) takes over for those frequencies where the electrolytic will not work properly. In short, these two capacitors *complement* each other, permitting both high-frequency and low-frequency operation of the circuit.

The output signal is taken pseudodifferentially. The 1K ohm load resistor is connected across the differential outputs of the MC1590 device, but one of the outputs (pin No. 7) is tied directly to the V+ line and is bypassed to ground. Presumedly, this tactic increases the voltage swing permissible in the output signal.

As in the case of the RCA circuit considered earlier, the Motorola MC1590 device is capable of controlled-gain operation. Pin No. 2 permits the gain of the circuit to be controlled. We connect pin No. 2 to an external voltage source through a 5.6K ohm resistor. If the stage is to run wide open at maximum gain, connect the resistor directly to the V+ power supply. But, if manual gain control operation is desired, connect it to a source of variable DC bias, such as the potentiometer arrangement of the CA3028 example. Automatic gain control potentials would also be connected to this point. In some receivers, the control of gain is more complex and may be a combination of manual and automatic gain control. The actual potential applied to the 5.6K ohm resistor, in that case, would be a summation of manual and automatic gain control voltages.

The circuit of Fig. 19-7, as well as some of the video amplifier circuits to follow, looks deceptively simple. But all can be a real hassle to build! At the frequencies that are processed in these stages, sloppy component selection or layout can result in some unanticipated results. Poor quality capacitors, for example, might not have the simple capacitive characteristic you might suspect. They will be complex amalgams of inductance, capacitance and series resistance. Resistors that are noisy will produce signals that will be amplified and passed to the output. In low-gain, low-frequency circuits, these noise potentials would not have any important effect on the output signal.

Another problem is proper layout. Inadvertant feedback signals, caused by poor layout or grounding practices, will couple sufficient energy back to the input to make the circuit oscillate. If there are no contrived resonances in the circuit, the device will oscillate at the natural oscillating frequency of the amplifier—usually several megahertz and maybe VHF.

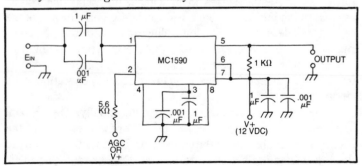

Fig. 19-7. MC1590 amplifier.

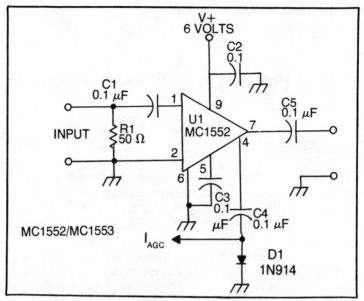

Fig. 19-8. MC1552 amplifier.

Another video amplifier is shown in Fig. 19-8. This circuit is based on either the MC1552 or MC1553 devices by Motorola Semiconductor Products, Inc. The differential input is configured such that one port is grounded. Resistor R1 is set to 50 ohms in order to match the transmission line inputs often required in video and other wideband amplifier applications. The circuit is pretty

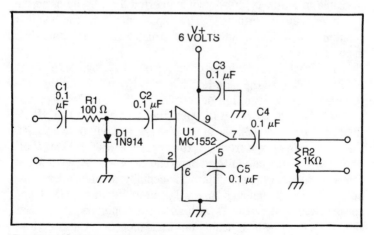

Fig. 19-9. MC1552 amplifier with level limiting.

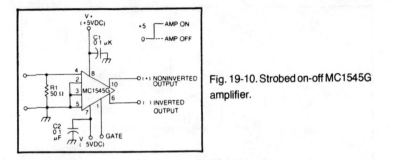

Fig. 19-10. Strobed on-off MC1545G amplifier.

straightforward and unremarkable, except for the use of agc . This circuit allows a small signal sample from the amplifier to be coupled to the agc rectifier, D1. The small potential developed across this diode is proportional to the strength of the signal, so it can be used to control the gain of other stages.

A similar video amplifier, also based upon the MC1552/1553, is shown in Fig. 19-9. This stage is functionally equivalent with that shown in Fig. 19-9, except that the input circuit is modified to provide frequency response tailoring (R1/C1) and overload protection (D1). The frequency response of this circuit can be arranged to produce a high-frequency rolloff at a frequency selected by the RC time constant of R1 and C1. The output is loaded by resistor R1, a 1K ohm unit.

The circuit of Fig. 19-10 is a little different: It is a unity-gain video switch. The circuit is based on the Motorola MC1545G device and features a differential output (single-ended output available in either inverted or noninverted form by ignoring the unwanted output terminal). The input circuit is the familiar configuration with a 50-ohm resistor to match the transmission line used as input cable.

The amplifier is switched on and off by a TTL logic level (0 and 5 volts) applied to the gate terminal (pin No. 1). When the gate terminal is LOW (logical-0, or zero VDC), the amplifier is off and will not pass a signal. But when the gate terminal is HIGH (logical-1, or +5 VDC), the amplifier is on and will pass a signal. The gain of the stage from DC to approximately 5 MHz is 0 dB (unity gain). The gain falls off approximately −6 dB at 10 MHz. These frequency response limits, incidentally, are dependent upon proper layout techniques and will suffer immensely if sloppy construction is allowed. As with all video amplifiers, the bypass capacitors must be high-quality units and must be located as close as possible to the case of the MC1545G IC.

Chapter 20
Using Integrators
and Differentiators

There are two operational amplifier circuits that have no substitute: differentiators and integrators. These circuits perform the electronic analog of the mathematics operations of differentiation and integration. The process of *differentiation* is the method by which we find the instantaneous rate of change of a function, in the case of the electronic circuit a voltage. The process of *integration* is inverse to the process of differentiation. Integration involves finding the area under a function (again, a voltage in the electronic case) between two limits. In this chapter, we will discuss both integrator and differentiator circuits. Included are some practical tips on circuit design and a couple of circuits that have been proven in actual service.

INTEGRATORS

An integrator must measure the *area* under a voltage-as-a-function-of-time curve. The simplest form of integrator circuit is shown in Fig. 20-1A and consists of a series resistor and a shunt capacitor. It is the charge in the capacitor that is proportional to the area underneath the E_{in} curve. The time constant of the RC network is selected to be longer than the period of the waveform being integrated.

Figure 20-1B shows an electronic integrator based on an operational amplifier—in other words, an active integrator circuit. In previous chapters, we derived the transfer functions of operational amplifier circuits using the properties of op amps, Ohm's law and Kirchoff's current law. The same process will be used to derive the transfer function of the operational amplifier integrator.

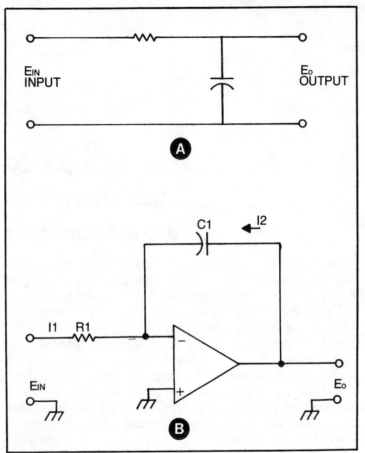

Fig. 20-1. RC integrator (A) and op amp integrator (B).

We know that

$$I1 = E_{in}/R1 \qquad \textbf{Equation 20-1}$$

and that

$$I2 = C_1(dE_o/dt) \qquad \textbf{Equation 20-2}$$

From the properties of the operational amplifier and from Kirchhoff's current law, we know that

$$I2 = -I1 \qquad \textbf{Equation 20-3}$$

By using the same substitution trick as before, we find that

$$C_1(dE_o/dt) = E_{in}/R1 \qquad \textbf{Equation 20-4}$$

218

Next integrate both sides of Equation 20-4:

$$\int \frac{C_1 dE_o}{dt}\, dt \;=\; -\int \frac{E_{in}}{R1}\, dt \qquad \textbf{Equation 20-5}$$

By collecting terms, E_o stands alone on the left side of the equal sign:

$$E_o \;=\; \frac{-1}{R_1 C_1} \int E_{in}\, dt \qquad \textbf{Equation 20-6}$$

Equation 20-6 is the transfer function of the operational amplifier integrators. Let's take a closer look at this equation to try to get some idea of how it works. The gain of the integrator is the factor, $-1/R_1 C_1$. The negative sign means that the polarity of the output will be opposite that of the input signal. If the input signal is positive, the output signal will be negative, and vice versa. But let's look at the denominator. How much gain does the integrator have? If capacitor C1 has a value of 100 pF and the resistor has a value of 10K ohms, the gain will be:

Table 20-1. Parts List for "Universal" Op Amp Integrator.

PART	DESCRIPTION	PART	DESCRIPTION
R1, R4, R6, R13, R15, R16, R18, R19,		C6	470 pf, silver mica or better
R21	10k	C7	100 pf, silver mica or better
R2	1k	C8	47 pf, silver mica or better
R3, R8	1 meg	X9, C10	200 μF, 25 WVDC or better
R5	2k, linear taper, 10 turn trimpot	C10	Same as C9.
R7	3.9 meg, 1% (or 5% metal film)		
R9	470k	A1, A3, A4	741 mini-DIP operational
R10	100k		amplifiers, or same as A2 for
R11	16 meg		superior performance.
R12	47k	A2	RCA CA3140T or CA3140AH.
R14	(same as R5)		Do not substitude with impunity.
R17	50k, linear taper, 10 turn panel		
	mount (1 turn usable if lower gain resolu-	S1	SPST miniature toggle
	tion is permissible, as it often is).	S2	SP11T rotary
R20	20k, otherwise sample as R17	S3	Normally open, pushbutton,
		SPST	
C1	0.1 μF, mylar or better	S4	SPST or DPDT miniature toggle.
C2	.02 μF, mylar or better		
C3	.01 μF, mylar or better	J1, J2	single-hole mounting BNC
C4	.005 μF, mylar or better		chassis connectors
C5	.001 μF, mylar or better	J3	Amphenol MS3102B-14S-5P
All resistors are 5% carbon film, 1/4 watt		J4	Mate of J3 in suitable or prefered
unless otherwise specified			shell (e.g., MS-3106B-14S-5S)

$$A_v = -1/R1C1$$
$$= -1(10^4 \text{ohms})(10^{-10} \text{farads})$$
$$= -1/10^{-6} = -1,000,000$$

The gain of the integrator with the component values specified is over 1,000,000! What does this mean in terms of the output signal? First, it means that the input signal is strictly limited to $E_{o(max)}/10^6$. If the maximum allowable output voltage is, say, 12 volts, then the maximum input voltage is $12/10^6$, or 12 microvolts! Another problem is that amplifier bias currents, normally so small as to be negligible, can become a real problem. If there is a gain of 1,000,000, normally tiny bias currents will charge the capacitor at a constant rate. The output voltage, then, will rise in a ramp-like fashion until it reaches the maximum output voltage allowable for that operational amplifier. This is a real problem, because it means that the integrator is available for only a short period of time—sometimes milliseconds.

Several things can be done to reduce the effects of high gain in an integrator. One of the first things to do is to select an operational amplifier that has a very high input impedance. The popular 741 operational amplifier is very cheap, but it will not work as an integrator if the gain is high. The input impedance is too low, so the input bias currents are too high. In an integrator developed by the author, it was found that the 741 device would saturate in only a few hundred milliseconds. The next selection was an $8 premium operational amplifier, the 725 device. This operational amplifier worked better than the 741 device, but it still saturated in short order. The time to full saturation was several thousand milliseconds instead of several hundred. It turned out that the best device for integrator service is the low-cost biFET operational amplifier made by RCA. One of these is the CA3140, while the other is the CA3160. These devices use a MOSFET transistor in the front end (input), and present an input impedance of 1.5×10^{12} ohms; that is 1.5 terraohms. This device worked well and saturated only very slowly. In some applications, an integrator made with such a device needs no operational circuitry.

Another tactic is to place an offset null circuit at the junction of the feedback capacitor and the input resistor. The purpose of this circuit is to insert a countercurrent that will counteract the action of the input bias currents of the operational amplifier. The potentiometer, however, must be very high resolution. Try one of the high resolution circuits presented in the earlier chapters.

The simplest case in which the integrator is used is when the input voltage is constant. The output voltage will ramp upward at a rate dependent upon the gain of the integrator and the amplitude of the input voltage. Consider a practical example.

☐ **Example:**

A constant DC potential of 2 volts is applied to the input of an operational amplifier integrator (Fig. 20-1B). Find the output potential at the end of 2 seconds if R1 = 500K ohms and C1 = 1 μF.

$$E_o = \frac{-1}{R_1 C_1} \int E_{in}\, dt$$

and, because E_{in} in this case is constant:

$$
\begin{aligned}
E_o &= \frac{-E_{in}}{R_1 C_1} \int dt \\[2mm]
&= \frac{(-2\,V)(t)}{(5 \times 10^5\,\text{ohms})(10^{-6}\,\text{farad})} \Big|_0^2 \\[2mm]
&= \frac{(-2)(2)}{(5 \times 10^{-1}\,\text{seconds})} = 8\ \text{volts}
\end{aligned}
$$

In other cases, the input waveform is not a constant DC level but will be some periodic function. Figure 20-2 shows the results of applying sinusoidal, triangle and square waveforms to the input of an integrator. In the case of the sine wave (Fig. 20-2A), the output, or bottom, waveform has the same shape as the input, or top, waveform, but is phase-shifted by 90 degrees. This is consistent with the results obtained when one mathematically integrates a sine wave. The tracing shown in Fig. 20-2B has a triangle at the input, and a sine wave at the output, This is in testimony that the integrator is basically a low-pass filter. It will strip off the harmonics of the triangle and leave only the sinusoidal output waveform.

The action of the integrator on the square wave is shown in Fig. 20-2C. When the input square wave is high, the capacitor will charge in a linear manner. The output will be nearly straight (some exponential shape is seen). When the phase of the input signal reverses, however, the capacitor will begin to discharge. If the input signal is truly constant and the integrator is perfect—which few are—the output will be a triangle waveform.

Practical Multirange Integrator Project

The circuit shown in Fig. 20-3A is a wide-range integrator which the author built for a physiologist whose research was on a

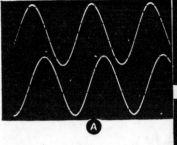

Fig. 20-2. Integrator waveforms: sine wave (A), triangle wave (B) and square wave (C).

low budget. Commercial integrator amplifiers cost $500 to $1000, yet this circuit worked just as well for the intended purposes. The circuit uses an RCA CA3140 biFET operational amplifier. A CA3160, which is a later type, will also work nicely. The integrator is amplifier A2. The drift cancellation, per the instructions given earlier in this chapter, is performed by selection of the CA3140 op amp and potentiometer R15, which serves to inject a countercurrent into the summation point of operational amplifier A2.

Another form of drift was not addressed earlier. If the applied signal has a DC component, a DC offset potential will tend to

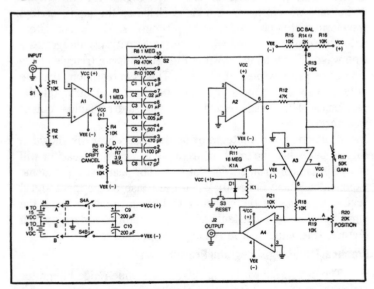

Fig. 20-3. "Universal" op amp integrator.

222

accumulate on the feedback capacitor. This charge is bled off by resistor R11. Note the extremely high value of resistor R11 (16 megohms). This value was selected as a trade-off between drift cancellation and deterioration of the integrator action. If this resistor is too small, the device will operate as an inverting follower amplifier with a 1 megohm input resistor and R11 as the feedback resistor. The capacitor that would normally integrate will then have the effect of reducing the frequency response of the circuit.

Integrator reset is provided by relay K1. When the operator presses the *reset* button on the front panel (S3), relay K1 will close, shorting out the integrator capacitor. This short circuit will discharge the capacitor and reset the integrator output to zero. Note that a diode is connected in parallel with the coil of relay K1. This is standard practice, especially in solid-state circuits. The coil of the relay is an inductor that will produce an *inductive kick* spike every time the reset button is operated. The pulse occurs on release of the switch.

Two principal factors in the selection of the capacitor value to use in the integrator are the period of the waveform being integrated and the maximum output voltage that can be tolerated. In general, the higher the input frequency is the lower the value that is required for the capacitor. At frequencies such as those encountered in my physiologist friend's research, the values ranged from 0.001 μF to 0.1 μF, with 0.01 μF being the most commonly used value. If too low a value is selected, the output of the circuit will clip rather badly. This is because of the gain problem mentioned earlier. In that case, the amplifier will saturate with low-amplitude signals, assuming that bias currents don't get it first. This is also the reason why an input attenuator was provided. In retrospect, it would have probably been better if an attenuating amplifier (an operational amplifier with a gain less than unity) had been provided. As it was, the need for an attenuator was not apparent until some months after the integrator was built and working.

The circuit shown in Fig. 20-3 has proven itself at frequencies between DC and 100 kHz. At higher frequencies, though, the 741 devices are not satisfactory, so additional CA3140/CA3160 devices should be used for the other operational amplifiers.

The rear end of the circuit consists of operational amplifiers A3 and A4. The master gain control, which also provides scaling of the output or the ability to adjust a strip chart recording to a specific span, is part of the circuitry in amplifier A3. The feedback resistor is a 50K ohm, linear taper, multiturn potentiometer. The gain is adjustable over the range of 0 to a little greater than unity. DC balance refers to keeping the base line stable when the capacitor range switch or the gain control is adjusted. This function is performed by an offset null potentiometer, R14. The output amplifier (A4) also incorporates an offset null-type control, but in this case it is an offset producing potentiometer that is used to provide position control.

Adjustment Procedure

Set the S1 and S4 open; S2 to position 11; R5, R14 and R20 to midrange; and R17 to maximum resistance. The procedure is as follows:

□ Close switch S4 and allow the circuit to warm up for 15 minutes. Do not become alarmed if the circuit appears to be saturated. It is, of course, but that is unimportant at this time.

□ Adjust potentiometer R20 for 0 VDC (± 10 mV) at point A.

□ Adjust potentiometer R14 for 0 VDC (± 10 mV) at point B.

□ Adjust potentiometer R5 for 0 VDC (± 10 mV) at point D.

□ Apply a 1V, 100-Hz (frequency is not too critical) square wave to jack J1.

□ Use an oscilloscope to check for output signal.

□ Disconnect the signal source. Turn switch S2 to position 8 and ground jack J1—the hot side!

□ Press reset switch S3.

□ If the oscilloscope trace of the output signal (scope sensitivity approximately 1 V/cm) shifts when the last step is performed, adjust R5 to cancel this drift.

□ Repeat the last two steps until no further improvement is noted. A slight amount of offset is permissible, provided that it forms a stable base line after S3 is released. This offset can be cancelled by the position control, so it is unimportant.

□ Reduce the resistance of R17 to zero. Then adjust it back to maximum.

□ If the base line shifts, adjust R14 to cancel the shift.

□ Repeat the last two steps until no further improvement is noted.

□ Apply a 10-Hz, 1V square wave to J1. Remove the input short.

□ The output waveform should be a triangle. If it is clipped, close S1 and reinspect the waveform. In low settings of S2, the gain of integration is high enough to cause clipping.

DIFFERENTIATORS

A differentiator is an electronic circuit that produces an output that is proportional to the instantaneous rate of change of the input signal. This value is called the *derivative,* or more properly, *first derivative,* of the input signal. The simplest form of differentiator is the RC circuit of Fig. 20-4A. Note that this circuit is just the opposite of the simple RC integrator shown earlier. You could almost guess this to be the case when you consider that the processes of integration and differentiation are inverses of each other; in fact, there is a fundamental theorem of calculus that tells us that we will obtain the function itself if we integrate the first derivative of the function. In other words, the process of integration would, in that case, reverse the process of differentiation.

A somewhat better method of differentiation is the circuit shown in Fig. 20-4B. This is an operational amplifier, or active, differentiator. Again, note the similarities and differences between this and the integrator circuit that uses the operational amplifier. Our analysis of this circuit will follow lines similar to the other analyses that have been performed. We know that

$$I2 = -I1$$
$$I1 = C(dE_{in}/dt)$$
$$I2 = E_o/R$$

so, by the same sort of substitution as yielded the answer before:

$$\frac{E_o}{R} = \frac{C\,dE_{in}}{dt}$$

solving for E_o:

$$E_o = \frac{R\,C\,dE_{in}}{dt}$$

Note that the RC time constant, which is the gain of the circuit, is in the numerator in this case. This is exactly the opposite of the situation found in the integrator circuit.

The time constant RC should be selected to be very short relative to the period of the waveform being differentiated. We don't even begin to see differentiator action until the time constant approximates one-tenth of the input waveform period.

Figure 20-5 shows what to expect when a square wave is applied to the input. Note that the leading edge has a very high positive rate of change. This causes the differentiator to produce a very high value output signal. Remember that it must be proportional to the rate of change of the input signal. When the positive top comes along, however, the output drops to zero (the exponential decay is a slight error caused by the necessity to discharge the capacitor and, possibly, the slew rate limitation of the operational amplifier). If the time constant of the differentiator is very small, it is likely that the output will be a series of positive and negative spikes, without the exponential decay portion.

The action of the differentiator on sine, square, and triangle waveforms is shown in Fig. 20-6. As in the case of the integrator, the operation of the differentiator on sine waves is to produce little more than a phase shift, but in the direction opposite that of the integrator. The inverse operation of the two circuits can be seen from this fact because, a sine wave is passed through an integrator, followed by a differentiator, the output sine wave would have been shifted forward 90 degrees, and then back 90 degrees, to the same phase it had when it entered the input circuit.

In Fig. 20-6B the same sort of waveform as before is seen. The square wave, when differentiated, becomes a series of positive and negative spikes. The triangle produces a square wave output (Fig. 20-6C). This is because both slopes of the triangle waveform have a *constant rate of change*. On the positive-going side, the output of the square wave is positive, reflecting the

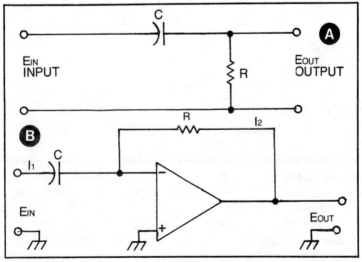

Fig. 20-4. RC differentiator (A) and op amp differentiator (B).

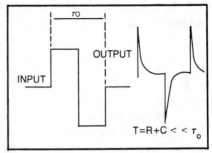

Fig. 20-5. Effect of differentiation on a square wave.

$T = R + C < < \tau_0$

direction of the change. Similarly, on the downslope of the triangle, the square wave output is negative.

Differentiator Problems

Like the integrator, the textbook differentiator, while appealingly simple, does not always work well in practice. The problem is the use of nonideal components. Perhaps the biggest problem is the frequency response of the operational amplifier. The integrator has little problem with this op amp property because it is a low-pass filter. But the differentiator is a high-pass filter, so it will have a rising response characteristic. Figure 20-7 shows an operational amplifier differentiator with a *fix* (R2) that will help overcome the problem of high-frequency oscillations. The Bode plot of the frequency response is shown in Fig. 20-8. Resistor R2 is placed in series with the input capacitor in order to reduce the natural frequency response of the operational amplifier. Figure

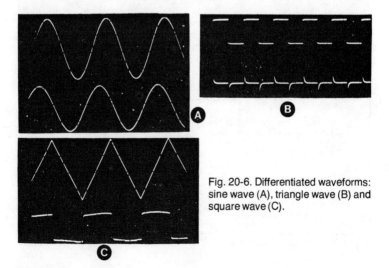

Fig. 20-6. Differentiated waveforms: sine wave (A), triangle wave (B) and square wave (C).

227

20-8 shows the normal plot of the operational amplifier frequency response and the response of the *ideal* differentiator. Note that the two curves intersect in the -12 dB/octave region of the operational amplifier response. This is potentially an unstable situation that can lead to high-frequency oscillation of the circuit. The effect of R2 is to modify the response of the differentiator so that it intersects the operational amplifier response at some frequency where the op amp gain is less than unity. This situation is inherently more stable than the open response version usually shown in textbooks. For most operational amplifier differentiators, the value of resistor R2 is 10 to 100 ohms, and it is found from:

$$R2 = 0.503/f_3 C_1$$

Capacitor C2 also helps to reduce the frequency response of the circuit, and its value is calculated from:

$$C2 = 1/(2\pi f_4 R_1)$$

Calibration of Differentiators

It is sometimes necessary to calibrate the differentiator so that its output can be quantified to reflect the actual rate of change of the input signal. One of the best methods is to employ a tactic used in many physiological electronic devices that include a differentiator. Build a linear ramp generator. A typical circuit might be an integrator connected such that its input sees a constant input voltage. Connect the linear ramp output of the generator to the input of the differentiator. Because the ramp has a constant rate of change, the output of the differentiator will take on a constant

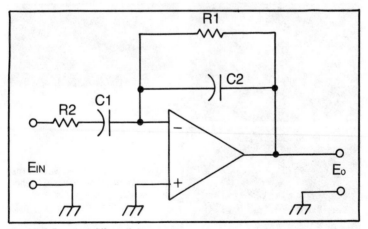

Fig. 20-7. Practical differentiator.

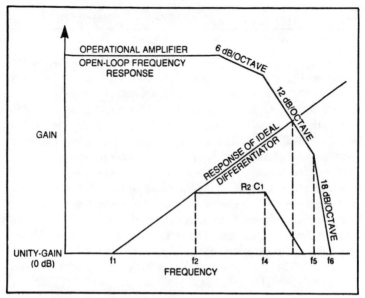

Fig. 20-8. Differentiator response curve.

value. It is then a relatively simple matter to determine the rate of change information required.

A differentiator can also be calibrated with a ramp or triangle generator *if* you know the slope of the waveform in volts/second. This value can be calculated or determined from examination of the waveform on an oscilloscope. Consider a 500-Hz triangle; it has a period of 2 milliseconds. If the output is adjusted to have an amplitude of 1V, the triangle will reach the 1V peak in one-half of its total period, or 1 millisecond. The slope, then, is:

$$(1V/1\,ms) \ \textbf{x} \ (1000\,ms/1\,s) \ = \ 1000\,volts/sec$$

We can then examine the square wave output of the differentiator for the voltage level; this level represents a slope of 1000 volts/second, or 1 mV/ms. It is then a simple matter to convert this measurement to the types of units that we are dealing with in the amplifier. Let us say, for example, that the differentiator is connected to the output of a blood pressure amplifier in a physiology laboratory. The units for pressure are millimeters of mercury (mm Hg), and a common scale factor is 10 mV/mm Hg. The output of the differentiator, in this case, will be dP/dt, or the time rate of change of blood pressure. The conversion to units of the input signal would be:

$$\frac{mm\,Hg}{10\,mV} \times \frac{1\,mV}{1\,ms} \times \frac{1000\,ms}{s} = \frac{100\,mm\,Hg}{s}$$

The applications of differentiators and integrators are too numerous to mention here. They are found extensive in electronic instrumentation that is used to process signals or to display some sort of numerical or waveform output. These circuits are often used to derive signals that are not easily "transduceable" in the real world. Consider the matter of velocity and acceleration in mechanics. Accelerometers and speedometers are somewhat harder to build than simple displacement meters. Velocity is the first derivative of displacement, however, while acceleration is the first derivative of acceleration and the second derivative of displacement. We would, in this case, build two differentiators and a displacement transducer. The transducer output would indicate displacement along some axis x. The output of the first differentiator would be the velocity $v = dx/dt$, while the output of the second differentiator would be acceleration $a = dv/dt = d^2x/dt^2$.

Practical Differentiator Project

The differentiator circuit shown in Fig. 20-9 was built to be the complement of the circuit shown earlier (integrator). This design was also created for a physiologist on a limited budget and takes the place of a commercial instrument that costs $400 to $1000, with little deterioration in performance for the purpose selected. This circuit has a multiple range selector switch, so it is actually quite versatile.

Note that the design of this differentiator is a little different from that of the classical differentiator shown in the earlier section of this chapter. The reason for this is that the other circuit was a little difficult to tame in some ranges. Despite the value of series snubber resistor selected, the darn thing would still oscillate at some settings of the range selector switch. The classical circuit works well when the range is fixed or limited to one of two carefully selected values of RC time constant. But when confronted with a large selection of ranges, problems develop. If a series resistor worked well on the short RC time constant, or high-frequency, ranges, it would fail to work on the long RC ranges. Similarly, a resistor large enough to move frequency f_4 low enough on the long RC ranges would make the circuit behave as a "low-low-low-pass" filter on the high ranges.

The answer was to use an RC differentiator circuit and then

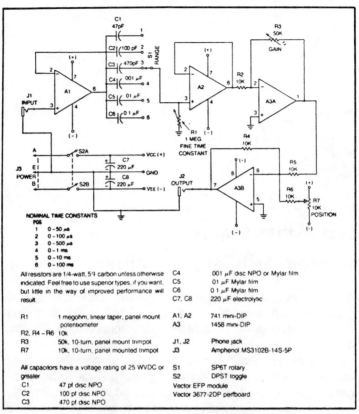

Fig. 20-9. "Universal" operational amplifier differentiator.

buffer it with operational amplifiers. This circuit, shown in Fig. 20-9, eliminates the problem of oscillation on all ranges of the switch, and all settings of R1. The actual differentiator network is potentiometer R1 plus capacitors C1 through C6. Amplifier A1 forms a input buffer amplifier that has unity gain. The second operational amplifier is also a unity-gain, noninverting follower, but it is an *output* buffer. Input operational amplifier A1 can be almost any low-grade operational amplifier that has a relatively low offset voltage and a low output impedance. Most 741 devices will work in this slot. The second operational amplifier, however, may require some type of premium device that has a high input impedance. The first model using 741 devices was successful, but in some other models, a biMOS or biFET input operational amplifier was needed. Amplifier A3 forms a standard "rear end" circuit that provides both sensitivity and position controls.

231

Chapter 21
Logarithmic and
Antilog Arithmic Amplifiers

An amplifier that has a logarithmic transfer function is used to compress the dynamic range of a signal. The transfer function of these amplifiers shows a saturation at high input voltages. This can be used to good advantage in many instrumentation applications and in some communications circuits. Sometimes, the range of analog data at the input of a circuit is greater than the ability of the circuit to handle it. At the high end, we find maximum allowable input voltages, while at the lower end there are limitations based upon the minimum amplitude signal that can be resolved. If the first amplifier is made a logarithmic stage, the range of the data is compressed to the range of the electronic circuit that processes the signal.

APPLICATIONS

The logarithmic amplifier can also be used for a number of signal processing and arithmetic operations. We can, for example, make analog modulators, or multipliers, and dividers using logarithmic amplifiers. Some of the circuits used for these functions would be quite complex were it not for log amps, as they are nicknamed.

But how does one go about making an amplifier that has a logarithmic transfer function? There must be an impedance in the feedback loop of a high-gain amplifier (operational amplifier) that has the proper logarithmic property. An ordinary pn junction has such a function. The volt-ampere characteristic of a pn junction diode is expressed by the equation:

$$I = I_o(e^{(V/nV_t)} - 1) \qquad \textbf{Equation 21-1}$$

where I is the forward current through the diode in amperes, I_o is the reverse saturation current of the diode (leakage) in amperes, e is the base of the natural logarithms, V is the applied potential in volts, n is a constant that depends upon the type of material used (approximately 1 for germanium diodes and 2 for silicon diodes), and V_t is the voltage equivalent of temperature, and is evaluated as kT/q, where k is Boltzmann's constant (1.38×10^{-23} J/°K), T is the temperature in degrees Kelvin (°K) and q is the electronic charge (1.6×10^{-19} coulombs).

Equation 21-1 evaluates to approximately $I_o(e^{V/nV_t})$ for those forward currents that are much greater than the reverse saturation leakage current. Because the reverse current tends to be in the 10^{-14} ampere range, it is highly likely in any practical circuits that I will be much, much greater than I_o, allowing the use of the simpler expression. It can be concluded, therefore, that placing a pn junction in the feedback loop will cause the amplifier to have a transfer function that is related to the logarithmic property of the diode.

In most cases, however, we obtain superior performance if we place a transistor in the feedback network, instead of an ordinary diode. This tactic is shown in Fig. 21-1, where an npn bipolar transistor (Q1) is connected in the feedback loop of an inverting follower operational amplifier. The base-emitter voltage of the transistor is a function of the collector current and leakage current, namely:

$$V_{be} = V_t \log((I_c/I_s) + 1) \qquad \textbf{Equation 21-2}$$

or, in the simpler form allowed when $I_c >> I_s$:

$$V_{be} = \frac{kT}{q} \log_e \left[\frac{I_c}{I_s} \right] \qquad \textbf{Equation 21-3}$$

where I_c is the collector current in amperes, I_s is the reverse saturation (c-e leakage) of the transistor, V_{be} is the base-emitter potential in volts, k is Boltzmann's constant (1.38×10^{-23} J/°K), T is the junction temperature in degrees Kelvin (°K), and q is the electronic charge (1.6×10^{-19} coulombs).

Equation 21-3 can be simplified a little if the assumption that the temperature will remain constant (it won't, but it's a start!) is made. Then evaluate Equation 21-3 partially:

$$V_{be} = \frac{(1.38 \times 10^{-23} \, \text{J/K})(300 \, ^\circ\text{K}) \log(I_c/I_s)}{(1.6 \times 10^{-19})} \quad \textbf{Equation 21-4}$$

$$V_{be} = 26 \, \text{mV} \log_e(I_c/I_s) \qquad \textbf{Equation 21-5}$$

This expression assumes room temperature to be 300° K, or 27° C. As it turns out, that is a fair approximation. We can see from Equation 21-5 that the base-emitter voltage is now dependent only on the collector current, the reverse current being nearly constant under the assumed condition of constant temperature. We are now ready to analyze the circuit in Fig. 21-1 for its operation. Use the same method as in the previous operational amplifier chapters and use the operational amplifier basic properties, Kirchhoff's current law and Ohm's law. We know, incidentally, that $V_o = V_{be}$ because of the circuit configuration.

$$V_{be} = V_o \qquad \textbf{Equation 21-6}$$

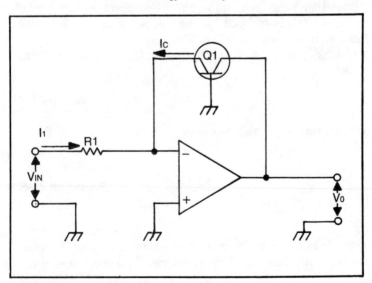

Fig. 21-1. Logarithmic amplifier.

and $$I1 = V_{in}/R_{in}$$ **Equation 21-7**

The properties of the operational amplifier tell us that
$$I1 = -I_c$$ **Equation 21-8**

so by substituting Equation 21-7 for I1, we obtain:
$$\frac{V_{in}}{R_{in}} = -I_c$$ **Equation 21-9**

Since $V_{be} = V_o$, we can write:
$$V_{be} = 26\,mV \log_e(I_c/I_s)$$ **Equation 21-10**

or
$$V_o = 26\,mV \log_e(I_c/I_s)$$ **Equation 21-11**

Now, substituting Equation 21-9 for I_c in Equation 21-11:

$$V_o = 26\,mV \log_e\left[\frac{V_{in}}{R1\,I_s}\right]$$ **Equation 21-12**

We know that R1 is always constant and I_s is reasonably constant given the assumption of constant temperature. Further, the logarithm of a constant is a constant, so Equation 21-12 is a transfer function that relates output voltage V_o to the logarithm of an input voltage. We are, therefore, satisfied that the circuit of Fig. 21-1 is a logarithmic amplifier.

We made an initial assumption that the temperature would remain constant in our log amp, but this is a highly unlikely situation unless we build the circuit inside a temperature-controlled oven. The term for the base-emitter voltage contains a direct proportionality to temperature, and this must be recognized in real circuits. One approach is to use a thermistor in a modified circuit, such as the one shown in Fig. 21-2. The temperature coefficient of the thermistor is selected to counteract the temperature dependence of the transistor in the feedback loop. This amplifier will track temperature over the normal room temperature range. We will shortly examine at least one other method of temperature stabilization. But first, let's look at a couple of other methods.

The equations thus far have emphasized the natural logarithmic scale, $\log_e$. Sometimes, however, we wish to use base-10 logarithms and would like to have a log amp with a base-10 property. Fortunately, the ratio of common (base-10) and natural (base-e) logarithms is a constant: 3.679. The output of the amplifier can be scaled with an times-3.679 amplifier stage to create an output that is dependent upon the common logarithm of the input voltage.

PRACTICAL CIRCUITS

A practical version of the circuit shown in Fig. 21-1 is shown in Fig. 21-3. This circuit was originally based on the 741 operational amplifier, but one is advised to consider some of the premium models that are now available at low cost. When this circuit was first presented, a pretty penny had to be paid for premium performance in an operational amplifier. But today, many companies are making operational amplifiers at low cost with specifications that were once regarded as premium. The lowly 741 is still around, but there is even evidence that its performance has improved considerably over the years.

The logarithmic element in this circuit is a type 2N2222 or 2N2218 npn silicon transistor. It is connected in the negative feedback loop of the operational amplifier. Some bias current will be present in this amplifier, and thus must be compensated. This

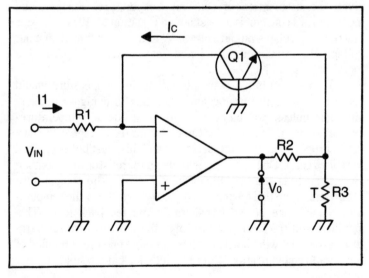

Fig. 21-2. Improved log amp.

236

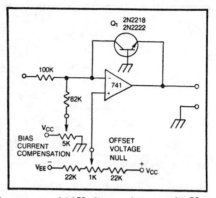

Fig. 21-3. Low-cost log amp.

has been done with a potentiometer and 82K ohm resistor to the V_{cc} positive power supply. The offset null potentiometer is used to calibrate the output voltage. Remember that the logarithm of zero is undefined, and that $\log 1 = 0$. When calibrating the circuit, you will input a voltage that you wish to define as one unit, and then vary the null for zero output voltage. The log amp is then calibrated over several decades of input voltage.

The inverse of the log amp is obtained by switching the respective roles of the transistor and resistor (Fig. 21-4). The input voltage becomes the collector voltage of the transistor, and the base-emitter voltage, V_{be}, becomes the differential voltage across the amplifier inputs. The transfer function of this amplifier produces an output that is the antilog of the input voltage. This allows us to decompress our analog data signal, or perform certain other functions that are dependent upon either "delogging" the signal, or producing the antilog of a voltage.

Figure 21-5 shows a practical logarithmic amplifier circuit that will operate over a range of at least four decades if 741-family operational amplifiers are used for A1 and A2. An input voltage in

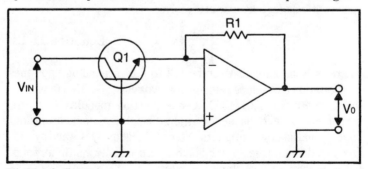

Fig. 21-4. Antilog amp.

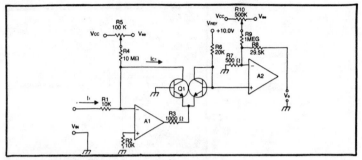

Fig. 21-5. Temperature compensated log amp.

the range of 2 millivolts to 20 volts can be used with 741 devices. Below approximately 2 millivolts, the input current will become almost the same magnitude as the transistor bias currents for the operational amplifier inputs. This situation can usually be improved by substituting a superior operational amplifier for A1 and A2. Note that some of the RCA biMOS and the biFET operational amplifiers will offer considerably lower input bias currents. The biMOS op amps have MOSFET input transistors and hence, very little input bias current. The CA3140 and CA3160 devices, for example, have input impedance values in the terrohm (10^{12} ohm) range, so the input bias currents would be picoamperes, or even femtoamperes (10^{-15} amperes). The use of these amplifiers in the circuit of Fig. 21-5 should reduce the lower end of the range significantly.

Logarithmic amplifiers are available in the form of analog function modules. The *Burr-Brown Research Corp.* manufacturers a line of these amplifiers. The Burr-Brown model 4127 is a straight log amp that will operate over a 120 dB range, if layout of the circuit is given due regard. Burr-Brown also offers a number of multifunction modules (discussed elsewhere in this book) that will allow a variety of transfer functions of the form:

$$V_o = K \left[\frac{V1}{V2} \right] \times V3 \qquad \textbf{Equation 21-13}$$

where m is a factor that can be 0.2 to 5.0, depending upon the configuration and values selected for two resistors. This device is the Burr-Brown 4301/4302. Other function modules that are essentially collections of logarithmic amplifiers include analog dividers and analog multipliers. Note also that many ICs and bybrid modules that are labeled "modulators" are actually multipliers and will often contain logarithmic amplifiers.

238

Chapter 22
Voltage-to-Current Converters

Voltage-to-current (E-to-I) converter amplifiers are often used in certain transducer applications and in those cases where the output display device will be a galvanometer or other form of analog meter movement. Some strip chart recorders are also designed to respond to current levels, rather than voltage levels. In this chapter, some of the basic circuits used to derive a current output from a voltage input will be covered.

THEORY OF OPERATION

Just what type of amplifier is used in a voltage-to-current converter? The transfer function of an E-to-I converter is of the form:

$$A = \frac{I_o}{E_i} \qquad \text{Equation 22-1}$$

where A is the gain of the amplifier, I_o is the output current of the amplifier, and E_i is the input voltage applied to the amplifier.

The units of the output-versus-input transfer function is *ampere/volts*, which is the reciprocal of resistance. Because conductance (G) is the reciprocal of resistance, a safe conclusion is that the current-to-voltage converter is a *transconductance amplifier*. Many different types of circuits can be used as a transconductance amplifier, but in this chapter the discussion is limited to operational amplifier circuits.

PRACTICAL CIRCUITS

One of the simplest circuits for E-to-I conversion is shown in Fig. 22-1. The circuit is a simple operational amplifier inverting follower in which the load resistor (through which the output current is made to flow) is used as the operational amplifier feedback resistor. Input voltage E_{in} sets up a current flow in resistor R_{in}. By Ohm's law, this current will be $I_i = E_{in}/R_{in}$. A second current, I_2, is set up in the load, or feedback, resistor. It will be equal to I_i (Kirchhoff's law and the basic properties of the operational amplifier). The output current, then, is equal to E_{in}/R_{in}, the same value as the input current. The output voltage, which would be important in almost any other application, is ignored in this case.

The circuit shown in Fig. 22-1 uses a floating output load. Another floating load E-to-I converter is the version shown in Fig. 22-2. This circuit is similar to the previous version in that it uses the inverting follower configuration of the operational amplifier. The load resistor (the output load for a current-output device) is series-connected between the operational amplifier output terminal and the feedback voltage divider. For purposes of analyzing the circuit, this can be considered as a voltage source driving a resistor network. Because these circuits have a transfer function that relates a change in output current to a change in the input voltage, it is a transconductance amplifier, which is exactly what is needed. The output current for the circuit of Fig. 22-2 is expressed by the equation:

$$I_L = \frac{E_{in}(R_L + R_f)}{R_{in}R_L} \qquad \textbf{Equation 22-2}$$

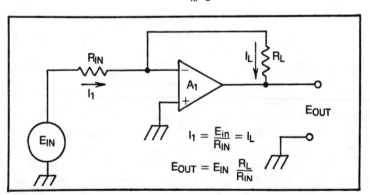

Fig. 22-1. E-to-I converter.

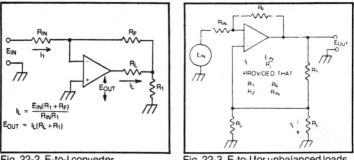

Fig. 22-2. E-to-I converter.   Fig. 22-3. E-to-I for unbalanced loads.

Please note that Equation 22 is one of the form required for a transconductance amplifier. We are, therefore, justified in using the circuit shown in Fig. 22-2.

The two circuits considered thus far have required a floating load. However, this situation is not always practical or possible. Indeed, the mind-set of most circuit designers is to use unbalanced, or grounded, loads. In those cases where the output load of the converter amplifier is to be grounded, a circuit such as shown in Fig. 22-3 will be required. This circuit will produce transconductance operation provided that the ratio, $(R1/R2) = (R_f/R_{in})$, in maintained. In that case, the output current will be:

$$I_L = \frac{-E_{in}}{R2} \qquad \textbf{Equation 22-3}$$

Note that this circuit depends not on absolute values of resistors, but only upon their *ratios*.

A number of other devices and circuits will provide the required transconductance operation. Some analog function modules, for example, will produce either a voltage output or current output, depending upon how the device is connected into the circuit. Some integrated circuits are also configured to produce a current output. In this same category, certain types of digital-to-analog converters (DACs) can be applied. The *multiplying DAC* uses an external voltage reference to produce an output current that is a function of both the voltage and a binary (digital) word applied to the inputs. The binary inputs can be used to provide either gain control operation, or an on-off switch, while applying the signal to the reference voltage input. Most of these devices will operate to frequencies of several kilohertz without introducing extremely large amounts of distortion. Nonmultiplying DACs, incidentally, contain an internal voltage reference source, so they cannot be used as amplifiers.

Chapter 23
Current Amplifiers

In Chapter 18, current difference amplifier (CDA) devices were covered. These are also called Norton amplifiers, for the well known Norton equivalent circuit used in analysis of electronic networks. These IC devices are indeed, current amplifiers, but they do not really cover the field properly. This chapter will cover those amplifiers that produce an *output voltage* that is proportional to an *input current*. These amplifiers are used frequently where a parameter is displayed on a voltage-sensitive instrument, such as an oscilloscope or strip chart recorder, even though the parameter is in the form of a varying current. Some of the basic types of current probes used in electronic measurement will also be considered.

CURRENT AMPLIFIERS WITH VOLTAGE OUTPUT

The transfer function of an amplifier is the ratio of the output function to the input function, and it describes the gain of the circuit. In the case of a generalized black box, the transfer function describes what the circuit does to the input signal. In the case of the current amplifier, the transfer function looks like:

$$A = \frac{E_{out}}{I_{in}} \qquad \text{Equation 23-1}$$

where A is the gain of the Circuit, E_{out} is the output voltage, and I_{in} is the input current.

What are the units of gain in this amplifier? Notice that the transfer function (Equation 23-1) describes the gain (A) in terms of a voltage divided by a current. By Ohm's law, volt/ampere is equivalent to the ohm—a *resistance*. As a result, the current-to-voltage converter is really only a *transresistance amplifier,* so A is more properly written A_r.

SOME SIMPLE CIRCUITS

The current amplifier must examine a current level flowing into the input line and convert it to an equivalent output voltage. There are a couple different ways this can be done. The simplest, which is the method used in many digital ohmmeters, is to cause the current to pass through a known resistance and then measure the voltage across that resistance. This is shown in Fig. 23-1.

The version shown in Fig. 23-1A uses a differential amplifier across the precision fixed resistor R. The output voltage will be expressed by:

$$E_{out} = I_{in} \times R \times A_d \qquad \text{Equation 23-2}$$

where E_{out} is the output voltage in millivolts, I_{in} is the input current in milliamperes, R is the input resistance, fixed and precise, and A_d is the differential gain of the amplifier.

☐ **Example:**

Find the output scale factor in millivolt/microamperes if input resistor R has a value of 10 ohms and the differential gain of the amplifier is times-100. (Hint: Assume a test current, such as 1 mA.)

$$
\begin{aligned}
E_{out} &= I_{in} R A_d \\
&= (1\,\text{mA})(10\,\text{ohms})(100) \\
&= 1000\,\text{mV a } 1\,\text{mA}
\end{aligned}
$$

The scale factor, then, is 1000 mV/mA. Now convert to microamperes.

$$\frac{1000\,\text{mV}}{\text{mA}} \times \frac{1\,\text{mA}}{1000\,\mu\text{A}} = \frac{1\,\text{mV}}{1\,\mu\text{A}}$$

The scale factor, then, can be expressed as 1 mV per μA.

This circuit is appealing in many cases, provided that the value of precision resistance R can be maintained at a low value. Higher values of resistance might tend to interfere with the accuracy of the measurement. The rule of thumb is that the value of the series resistor must be very much lower (less than one-tenth) the resistance of the circuit being measured.

The second alternative for this same idea is found in Fig. 23-1B. In this case, the amplifier is single-ended and is in the form of a noninverting follower with gain. The output voltage is expressed by:

$$E_{out} = \left\{ \frac{R1}{R2} + 1 \right\} \times I_{in} \times R \qquad \text{Equation 23-3}$$

where E_{out} is the output voltage, R1 is the feedback resistance, R2 is the "input" resistance, R is the resistance through which the current flows, and I_{in} is the input current.

Examples using this circuit are pretty much the same as in the previous example, except that the gain (A_d) is expressed by the quantity in brackets. You should recognize this expression as the ordinary equation for the gain of a noninverting follower using an operational amplifier device.

Several problems plague this particular design. One is the fact that we are limited to relatively low values of current, unless a pretty heavy wattage resistor is used, in which case, accuracy suffers a little—*precision* power resistors are a little hard to find. Another problem is that at low current levels, the bias currents of the operational amplifier produce as much output as the signal current. The bias currents of a 741, for example, can be up to several milliamperes. This offset voltage can be nulled out, but that does little for the effect of the input bias currents on the circuit being measured. To prevent this problem, it is necessary to use a high-input impedance operational amplifier. The RCA CA3140/CA3160 devices, for example, will provide input impedance levels of 10^{12} ohms. The CA3140 is listed as 1.5 terraohms. In many cases, these circuits are the circuits of choice.

Another approach to this problem is shown in Fig. 23-2. In this circuit, input current I_{in} is applied to the inverting input of an operational amplifier that is connected as an inverting follower. The output voltage is expressed by

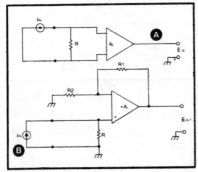
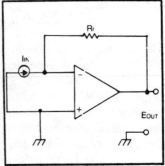

Fig. 23-1. I-to-E converters. Simpler circuit is at A. Fig. 23-2. I-to-E converter.

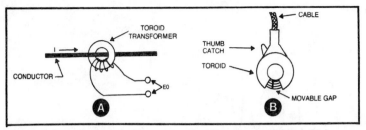

Fig. 23-3. Simplified circuit of a current probe (A) and its construction (B).

$$E_{out} = I_{in}R_f \qquad \textbf{Equation 23-4}$$

where E_{out} is the output voltage in volts, I_{in} is the input current in amperes, and R_f is the feedback resistance in ohms.

☐ **Example:**

A current of 500 microamperes flows into the input. Find the output voltage if the feedback resistance is 10K ohms.

$$
\begin{aligned}
E_{out} &= I_{in}R_f \\
&= (5 \times 10^{-4} \text{ amperes})(1 \times 10^4 \text{ ohms}) \\
&= 5 \text{ volts}
\end{aligned}
$$

The scale factor of this circuit, then, would be 5 volts/500 μA, or 50 mV/μA. We could then calibrate an output meter in terms of current, even though it was measuring units of voltage. In the previous example, a 0V to 10V digital panel meter would read 5.00 volts for 500 μA of current.

CURRENT PROBES

The circuits presented thus far require that the circuit be broken before a current measurement is made. Both circuits suffer from this need and from the possibility that the process of measurement might change the actual current. This is the current equivalent of circuit loading by a low-impedance voltmeter, which should be familiar to many readers. The probe in Fig. 23-3 will allow the indirect measurement of AC currents from 100 Hz to several megahertz. The probe consists of a toroidal inductor coaxial to the current-carrying wire. This is shown schematically in Fig. 23-3A. The output signal is a voltage that is proportional to the applied current.

The physical form of the probe is shown in Fig. 23-3B. The coil is wound on a toroid inside of a plastic housing. A movable magnetic gap, operated by a thumb catch, allows the probe to be slipped over the wire. The magnetic field created by the moving electrical charge in the wire produces a magnetic induction in the probe, thereby producing an output voltage.

Chapter 24
Chopper, Carrier
and Lock-In Amplifiers

Many people who work in engineering and scientific laboratories have a need for some very specialized amplifiers. Several types have become almost standard over the years: *chopper, carrier* and *lock-in.* The chopper amplifier is used to process certain low-frequency, low-amplitude signals that would otherwise be difficult to handle in DC or AC amplifiers with a low-end frequency response above 50 Hz or so. The carrier amplifier supplies an alternating current signal to an external circuit, such as a Wheatstone bridge transducer and then amplifies the portion of the AC signal returned to the amplifier input. These amplifiers are used to excite transducers that cannot handle DC. They are also used to amplify the output of certain types of DC transducers that seem to work better when AC-excited. One reason for this is that the AC amplifier used as the carrier amplifier can be controlled much more tightly with degenerative feedback and is therefore far more stable. The lock-in amplifier is a special type of carrier amplifier in which synchronous circuit techniques are used to bring in signals that would otherwise be lower than the noise level. These signals are impossible to process in ordinary amplifiers.

CHOPPER AMPLIFIERS

DC amplifiers have a tendency to drift, and they may be noisy. Neither of these factors is important when the amplifier gain is low, or even medium range, but each becomes critical in high gain applications. When the gain of the amplifier reaches times-500 or times-1000, these factors become somewhat more important.

Consider a hypothetical case where the drift is 10 μV/°C. If the amplifier gain is times-100, the output voltage change due to drift is:

$$(10 \, \mu V/°C) \times (100) = 1 \, mV/°C$$

But what happens in an amplifier that is intended to process low-level biological signals, such as the human electroencephalograph? These amplifiers might easily reach gains in the times-10,000 to times-100,000 range. In the latter case, the output drift due to a first-stage gain of 10 μV/°C would be:

$$(10 \, \mu V/°C) \times (10^5) = 1 \, V/°C$$

Figures like 1V don't seem to be very large until you consider a few simple facts about such systems. One is that the oscilloscope/strip chart recorder inputs used to display these signals will be 1V or 10V inputs. In the latter case, the drift is 100 percent of the input sensitivity! Let us assume that we are using a 10V input, and the drift is 1 V/°C. A 10° C change in the ambient temperature will make the amplifier saturate. This is the reason why we meed an amplifier with a low drift factor in the input stage. Later stages, incidentally, can be a little worse in the drift department, because they are not followed by as much gain as the first stage.

Noise is also a factor in high-gain applications. Noise in low-gain and medium-gain amplifiers is of little concern because the noise amplitude at the input is low amplitude relative to the signal amplitude. But in high-gain applications, where very low signal levels are processed, the noise rms amplitude may easily be equal to the signal amplitude. Amplifier noise is usually measured in terms of *nanovolts of noise per square root hertz* ($\overline{E}_n = nV(Hz)^{1/2}$). A typical bargain-grade amplifier will have a noise figure of 100 $nV(Hz)^{1/2}$. If the bandwidth of the amplifier used to process this signal is 10 kHz, the input signal will contain a noise component with an amplitude of:

$$\overline{E}_n \begin{aligned} &= (100 \, nV)(10^4)^{1/2} \\ &= (100 \, nV)(100) \\ &= 10,000 \, nV = 10 \, \mu V \end{aligned}$$

In a system where the gain of the amplifier is times-10,000, the output signal noise component will be equal to 100 mV, which could easily be 10 percent or more of the total output signal.

The *chopper amplifier* helps reduce both the noise artifact and the drift artifact in such circuits. Most low-level transducer, or biological, amplifiers are choppers; all medical amplifiers are chopper models.

The drift problem is cured by two properties of the AC amplifier used for the chopper circuit. One is the inability of the AC amplifier to pass signals near DC. The drift signal component is almost inevidibly a low frequency varying DC level. The second property is the use of heavy doses of degenerative feedback to stabilize the AC amplifier. The noise problem is cured by the use of a narrow bandwidth around the chopper frequency of the amplifier.

The low-frequency AC signals (above the drift rate) that many of these amplifiers process must be able to pass through the amplifier. They will not normally pass through such an amplifier that has a low-frequency response that is so much higher than their own frequency range. The solution is to *chop* the signal at a frequency that will pass through the amplifier. The chopped AC signal can then pass through the AC amplifier. It can then be demodulated at the output to recover the original waveform at a higher amplitude.

Figure 24-1 shows a basic chopper circuit. The most traditional type of chopper, which is still used in new equipment, consists of a vibrator-driven single-pole-double-throw switch (S1) that is connected into the circuit such that it alternately grounds out first the input and then the output terminals of the amplifier. An example of a chopped waveform is shown in Fig. 24-2. A low-pass filter must be used following the chopper stage in order to eliminate the chopper hash and any miscellaneous noise products that survive the demodulation process. Usually, the low-pass filter is one of the last stages in the amplifier.

Most mechanical choppers have a fixed rate of vibration, and this is marked on the case. Typical chopper frequencies are 60 Hz, 100 Hz, 200 Hz, 400 Hz and 500 Hz, with the 60-Hz and 400-Hz frequencies being most common. The main criterion for selecting a chopper frequency is that it be at least *twice* the highest frequency contained in the input signal. Note that this not the fundamental frequency of the input signal, but the highest frequency *component*—the Fourier series of the input signal.

A differential chopper amplifier is shown in Fig. 23-3. This circuit is probably the most common in actual equipment, possibly because the same signals that would benefit from the use of a chopper amplifier would also benefit much from the use of a differential amplifier, such as biological signals and transducer signals used in noisy environments. The input circuit for this amplifier is a center-tapped audio transformer. One input terminal of the amplifier is connected to the primary center tap of the input transformer. The other input terminal is switched back and forth between the ends of the transformer primary by the chopper switch.

The chopped AC signal can be amplified, but is not suitable for output unless the original unchopped waveform can be recovered. This is done by a synchronous demodulator following the AC amplifier. This stage is sometimes close to the output amplifier, while in others, the demodulation job is done relatively early, and then the following stages are DC amplifiers of low gain or medium gain.

Most chopper amplifiers that have been in service for any length of time are *mechanical-vibrator* models. The chopper switch is a vibrator-excited switch. Modern amplifiers of this category are a little more sophisticated, however, and will use CMOS or JFET electronic switches for the chopper. Several semiconductor manufacturers offer electronic switch IC packages that will do the job (even the CD4066/CD4016 will work), while others make IC devices that are specifically selected for chopper service. Electronic switching allows somewhat better control, less hash caused by switch contact noise and certain other advantages. They also don't wear out the contacts. Some manufacturers of hybrid or monolithic IC chopper amplifier devices, which could form the front ends of the type of amplifier that we are discussing, use a special type of *varactor bridge* as the chopper element. Similarly, some have used *optoisolators* for the chopper. In this case, the drive signal is applied to the LED, while the photodiode or phototransistor inside the isolator is used as the electronic switch.

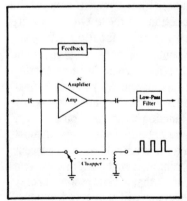

Fig. 24-1. Simple chopper amplifier block diagram.

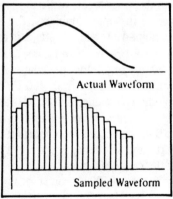

Fig. 24-2. Sampled waveform.

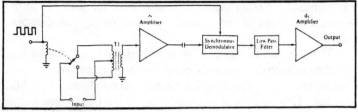

Fig. 24-3. Differential chopper amplifier.

The chopper amplifier limits the output noise by two methods. One is the low-pass close to the output. The other is that the AC amplifier can have a frequency response that is narrow and centered around the chopper frequency.

AC CARRIER AMPLIFIERS

A *carrier amplifier* is a signal-processing amplifier in which the signal that carries the desired information is modulated onto another signal; hence, it is analogous to a radio *carrier* signal. There are two types of carrier amplifiers. These are classified according to excitation method: DC and AC.

An example of the DC-excited carrier amplifier is shown in Fig. 24-4. A Wheatstone bridge transducer is used as the signal source, and it is DC-excited. This means that the potential applied to the transducer is a regulated, precision DC source. The transducer output is proportional to the excitation potential and the value of an applied stimuli (temperature, pressure, force, displacement, etc.). This signal, then, is a low-level DC signal that is proportional to the applied stimulus. The transducer signal is usually very noisy in addition to being low level. An amplifier builds up its level, while the low-pass filter trims off some of the noise. In some amplifiers, the functions of these two stages are combined into a single stage; in other words, it is a filter with gain.

The preprocessed signal at the output of the filter stage is then used to amplitude modulate a carrier signal. Typical carrier frequencies range from 400 Hz to as high as 25 kHz. The most common frequencies are 1 kHz and 2.5 kHz. The signal frequency response of a carrier amplifier is a function of the carrier frequency, so this can become an important parameter in some cases. In most texts, the frequency response of the carrier amplifier is 25 percent of the carrier frequency. A carrier amplifier that uses the low-end frequency of 400 Hz can process signals that have components only up to 100 Hz. A 25-kHz carrier frequency will support a frequency response of 6250 Hz. The AC amplifier following the amplitude modulator will merely add amplitude to the signal.

250

The *phase-sensitive detector* (PSD) demodulates the amplified AC signal. Envelope detectors are used in some cheaper models, but these are undesirable because they suffer from an inability to distinquish between real signals and spurious signals. An example of a phase-sensitive detector is shown in Fig. 24-5. Transistors Q1 and Q2 are used as electronic switches that provide a return path to ground for opposite ends of the secondary of transformer T1. These transistors are alternately switched on and off (out of phase with each other) by the reference signal such that Q1 is turned on hard when Q2 is turned off, and vice versa. The output signal from the PSD is a full-wave rectified version of the input signal. This signal can be passed through a low-pass filter to remove the residual AC which would be *ripple* in power circuits, to obtain the DC output waveform desired.

There are other methods of building PSD circuits than the use of bipolar transistor switches. Several other types of electronic switching, including either JFET or MOSFET transistors or IC switches that are based on these types of transistors (CD4016/CD4066, etc.) exist. Any type of double-pole-double-throw switch should work, and even diodes have been pressed into service. The PSD switches are toggled by the reference signal, so they will operate at that frequency. The idea is to toggle the switches in such a way that the output of the amplifier is always positive-going, regardless of the phase of the input signal.

The phase-sensitive detector is considered better than the simple envelope detector because it will reject all frequencies that are not close to the carrier frequency and those signals that are at the carrier frequency but do not have the correct quadrature

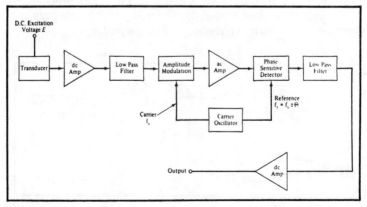

Fig. 24-4. Carrier amplifier.

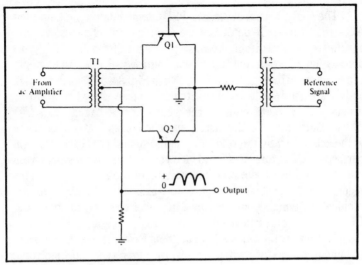

Fig. 24-5. Phase-sensitive detector.

relationship to the actual carrier. The PSD will also reject the even harmonics of the carrier itself.

The carrier amplifier will, however, respond to the odd harmonics of the carrier signal, and this can be a problem. Some manufacturers seem to ignore this problem altogether or try to keep the linearity of the carrier signal so high that harmonics are not a problem. In other cases, though, including many of the highest quality devices on the market, the AC amplifier is designed to have a frequency response limited to the carrier plus or minus 25 percent of the carrier frequency. This selection of frequency response limits will eliminate the third- and higher order harmonics before they reach the PSD. The purity of the output can be assumed by keeping the reference oscillator signal pure.

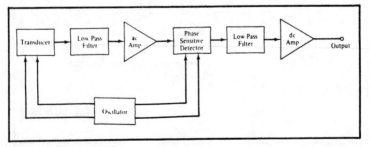

Fig. 24-6. Carrier amplifier.

The most common form of carrier amplifier is the AC-excited form shown in Fig. 24-6. In this type of circuit, the transducer or input circuit is AC-excited by the carrier signal. This tactic eliminates the need for the amplitude modulator. The use of AC excitation allows certain types of inductive and capacitive transducer to be used that are considerably more stable than the resistive strain gages used with DC amplifiers. Certain medical blood pressure amplifiers based on resistive techniques would drift completely out of calibration when someone handled the transducer. Heat from the operator's hand would cause the resistance elements inside the transducer to change.

In the circuit of Fig. 24-6, the small signal from the output of the transducer is amplified and filtered before being applied to the input of the phase sensitive detector. Again, some designs use a bandpass AC amplifier to eliminate odd-harmonic response. This circuit allows cancellation of transducer offset errors in the PSD circuit, instead of the transducer, by varying the phase of the reference signal supplied to the PSD circuit. Amplifiers following the PSD circuit can be DC amplifiers.

LOCK-IN AMPLIFIERS

Many of the amplifiers discussed thus far will produce relatively huge amounts of noise and will respond to any noise that is present in the input signal. This is quite common in transducer amplifiers, or where biological signals are being processed. The noise amplitude at the output is directly proportional to the square

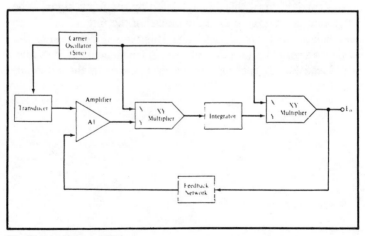

Fig. 24-7. Lock-in amplifier.

root of the circuit bandwidth. The lock-in amplifier is a specialized carrier amplifier in which the bandwidth is very narrow. Many lock-in amplifiers will use the circuit of Fig. 24-6 but will limit the overall frequency response of the amplifier by narrowing the bandwidth of the AC amplifier stages. The lock-in amplifier, in its most basic form is then a high-Q carrier amplifier.

Commercial carrier/lock-in amplifiers are available with carrier frequencies anywhere between 1 Hz and 200 kHz. In the case of the 1-Hz version, only the slowest-changing signals are processed. The lock-in amplifier principle works because the information signal is made to contain the carrier frequency in a way that is easy to demodulate and interpret. The AC amplifier accepts only a narrow band of frequencies centered around the carrier frequency. The narrow bandwidth frequency response permits greatly improved signal-to-noise ratio, but it also limits the lock-in amplifier to very low-frequency input signals. Even then, it is sometimes necessary to time-average the signals for several seconds to obtain the needed data.

Lock-in amplifiers are capable of reducing the apparent noise and retrieving signals that would otherwise be buried in a high noise level. Some models are able to offer improvements up to 100 dB, with 85 dB being relatively easy to obtain. The higher degree of noise suppression, incidentally, mades the cost go up considerably.

A somewhat more sophisticated form of lock-in amplifier is shown in Fig. 24-7: the *autocorrelation amplifier*. This amplifier uses a carrier signal that is modulated by the input signal. The composite is integrated, and the output of the integrator is then demodulated in a product detector. This type of lock-in amplifier produces a very high output at the proper frequency, but the output is reduced for signals that are not in phase with the reference signal.

Chapter 25
Problems with
Applying Amplifiers

There are sometimes problems in the application of amplifiers, and some of these almost defy rational answer. Some of these problems were covered under the rubric *operational amplifier problems* in Chapter 14. Although we were being specific to operational amplifiers, some of these are seen in all amplifier applications regardless of the type.

MISCELLANEOUS NOISE

A number of different types of noise will affect the operation of amplifiers. Some of these are man-made noise sources, while others are natural and are part and parcel of the design of the amplifier. It is hoped that the designer of an amplifier will take strategies that will reduce this type of noise.

The tactic that will be appropriate for suppression of external noise will, in large measure, depend upon the frequency of the noise. If the noise is of mostly high-frequency nature, we may often filter out the noise. We can, for example, use either a low-pass filter somewhere in the amplifier or custom tailor the frequency response of the amplifier to roll off the higher frequencies. Some commercial laboratory amplifiers have a switch-selectable frequency response that allows the user to set the upper and lower -3 dB points on the response curve.

If the noise is rf in nature, such as interference from a nearby radio transmitter or other source of high-power rf energy, you might use a combination of amplifier shielding and filtering. An L-section filter can usually be connected in series with the input leads of an amplifier without doing great harm to the frequency response—unless it too contains part of the rf spectrum.

Figure 25-1 shows how LC filters might be used in the inputs of an amplifier. The filter inductors should be connected right to the input jacks to minimize the length of unfiltered cable inside the shielded compartment. Note also that the filter network is shielded from the *outside world and from the amplifier.* In most cases, this means a small shielded compartment for the input filter inside the shielded cabinet of the amplifier.

Impulse noise is made up of high-frequency Fourier components, giving the noise pulses a characteristic fast rise time. But the fundamental frequency of these pulses is often quite low—on the order of a few kilohertz or less. The fundamental frequency is often coincident with the frequency band required of the amplifier.

Filtering is not often a good way to handle impulse noise because the action of the filter on the pulses is to *broaden* the pulse. Of course, a broadened pulse is worse than a narrow pulse. We can, however, take some pages from the notebooks of communications receiver designers: Noise blankers can be used. A *noise blanker* is a circuit that senses the noise pulse and then turns off the amplifier for the brief instant. The output signal will contain a hole, rather than a noise pulse, but this is generally less critical than the noise. Shielding, however, also benefits the system subjected to impulse noise. Try adequate shielding before trying one of the other methods.

One of the most common forms of noise experienced in electronic amplifier systems is 60-Hz pickup. The power mains in the U.S. and Canada operate at a frequency of 60 Hz. We find, therefore, that 60-Hz fields can be found almost everywhere.

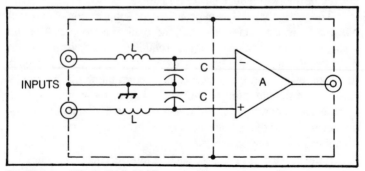

Fig. 25-1. Input filtering and shielding.

These fields will cut across input signal lines and cause interfering signals at the amplifier. In audio amplifiers, the 60-Hz signal appears as a hum in the output. In instrumentation circuits, it may well show up as some deviant behavior that looks like anything from a wiggle on the base line of a tracing or scope display to something a lot more malignant.

Applications where signal levels are low or where the signal source is located a long way from the amplifier input may well require a *differential amplifier* because of the inherant suppression of common-mode signals (see Chapter 12). In most cases—not all—the 60-Hz field will affect both inputs of the differential amplifier equally. This makes the output almost totally free of any 60-Hz signal, because of the common-mode suppression found in those types of amplifiers.

As mentioned, it is not always true that the 60-Hz input voltages are equal. It is sometimes possible to manufacture a differential signal from a signal that started out as common mode. There are two basic phenomenon that cause this problem, and both are essentially due to the *improper* use of *shielding*.

One of the major problems is the matter of the *ground loop* (Figs. 25-2 and 25-3). Figure 25-2 shows a differential amplifier connected to signal source E_{in} through a shielded cable. The problem in this example results from the use of *too many* grounds. Note that the shielded input signal source, each input line and the DC power supply are all grounded to different points on the ground plane. Let us say that DC power-supply current I passes along the ground plane to the common terminal on the amplifier. There will, unfortunately, be several voltage drops along this path due to the ground resistance. This resistance is not very large, but it can create a voltage drop that will be seen as a valid input signal by the amplifier. The voltage levels might be small, but so are some of the signals. If the gain is large and the amplifier is a power amplifier, the problem is exacerbated even more!

The main solution to this problem is a single-point grounding, as shown in Fig. 25-3. The grounds for the different parts of the system are connected together at a single point on the ground plane. This will eliminate some of the spurious voltage drops in the system. Frequently, it is necessary to try several different ground configurations before success is found.

Sometimes, amplifier systems have several subgrounds, and these are tied together at some common point. The common point is often a point on the chassis or one of the printed circuit board

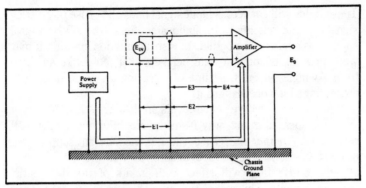

Fig. 25-2. Scenario for disaster.

edge connector pins. Examples of where this might be found include systems were there are high-level and low-level analog signals mixed on the same PC card, or analog and digital signals mixed on the same card. When there are large current and voltage transitions that are normal to digital circuits, ground loop problems are sometimes found in the analog circuitry associated with the digital. Many analog-to-digital and digital-to-analog converter (IC and hybrid) manufacturers even bring separate pins to the package body for analog and digital grounding!

In some cases, it is expedient to leave the shield open at one end of its run! Does that surprise you? There are a number of cases, especially when dealing with low-level, low-frequency, signals, when better performance is obtained with the shield open-circuited at the source end.

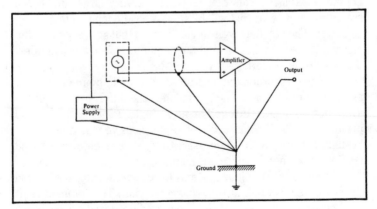

Fig. 25-3. A cure for Fig. 25-2.

HUM

Hum problems can also be reduced if the analog signal line from the source to the input is made a low impedance. The development of interfering voltage will be less in low-impedance circuits than in high-impedance circuits. Most amplifiers have a high-impedance input, as do many sources. If the signal is audio, or some other reasonable frequency AC signal, use a simple transformer. If it is otherwise perhaps a DC signal use a buffer amplifier at the source end. This is one of the uses of the unity-gain noninverting follower amplifier discussed in an earlier chapter.

It is also a good idea to amplify the signal before applying it to the line. The noise pickup of the line will be constant, regardless of the signal amplitude applied. Say that you have a 10 mV AC hum component and a 20 mV signal. The signal-to-noise (S/R) ratio, then, is 2:1. But what if the signal were amplified by 10, or 100? The S/R ratio would then be 200/10 (20:1) or 2000:10 (200:1)!

Audio lines in radio stations are often unshielded, yet experience no interference. How is this possible? A combination of techniques result in proper operation. One is to always drive long lines with a buffer amplifier, as just discussed. This will keep the signal level high and the relative hum level low. Another technique is to use a balanced 600-ohm line (Fig. 25-4). The audio source (microphone) is connected to a balanced pair, meaning both lines are above ground, through a line transformer T1. This transformer will convert the impedance of the source to 600 ohms. At the other end of the line, a similar transformer accepts the 600-ohm balanced line and applies the secondary signal to the input of the amplifier.

Figure 25-5 shows some of the causes and cures of differential signals manufactured from common-mode signal sources. Figure 25-5 shows the ordinary shielded cable system, with its equivalent circuit at Fig. 25-5B. Differential signal E is applied across the inputs of the amplifier, while a common-mode maybe hum—signal,

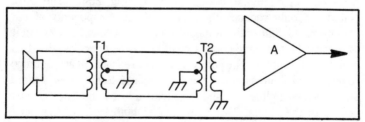

Fig. 25-4. Balanced line AC input.

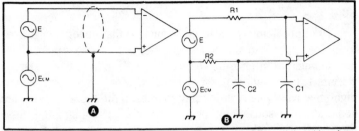

Fig. 25-5. How to make a common-mode signal act as a differential at either A or B.

E_{cm}, is connected from the inputs to ground. Resistances R1 and R2 are the source resistances plus the line resistances (which can be substantial in long runs). The capacitances are the capacitances of the two cables; all shielded cables produce some capacitance, and it can reach subtantial levels in cables over 50 feet long. The system works well only if the circuit remains *balanced,* meaning R1=R2 and C1=C2. Unfortunately, some very small unbalances will cause major problems, especially in high-gain cases. The differential signal arises from the fact that R1C1 $\neq$ R2C2, so $E_{C1} \neq E_{C2}$. The differential signal is $E_{C1} - E_{C2}$, where both components are manufactured from the common-mode signal, E_{CM}.

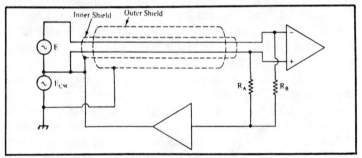

Fig. 25-6. Optional common-mode amplifier.

The manufactured differential signal can be eliminated in many cases by the use of a guard shield (Fig. 25-6). The extra shield not only reduces the values of the components of the manufactured differential voltage (reduced values of C1/C2), but it also cancels the remaining signal. This is done by the two resistors, R_A and R_B. The components from each input are summed together and applied to the inner guard shield. Sometimes, a guard shield drive amplifier is used to provide a low-impedance driving source to the shield. This tactic places both sides of the cable capacitances at the same potential, so $E_{C1} = E_{C2} = 0$.

Chapter 26
General Oscillator Theory

An oscillator is a circuit that will produce a periodic, or repetitive waveform such as a sawtooth, square wave, triangle wave, or sine wave. There are two basic forms of oscillator circuits: relaxation oscillator and feedback oscillator. The *relaxation oscillator* depends upon negative resistance devices. These devices will behave in the manner dictated by Ohm's law at potentials below some threshold voltage. The current will increase proportional to the applied voltage in that region. But after the threshold is exceeded, the current will *decrease* for increasing voltages. Other relaxation oscillators depend upon a related phenomenon in which the current is zero below some threshold potential and then increases to a very large amount at potentials above the threshold voltage. The neon glow lamp and unijunction transistor (UJT) fit into this category. The *feedback oscillator* depends upon the properties of the amplifier and feedback loop in order to oscillate. These oscillators make up a large segment of the electronics circuits that generate sinusoidal waves. In this chapter, both forms of oscillator circuit will be considered.

NEON LAMP RELAXATION OSCILLATORS

The neon glow lamp is a special form of lamp that uses ionized gas inside a glass envelope to produce light. Two electrodes are inside the glow lamp. When the potential across these electrodes exceeds the ionization threshold potential, E_t, the gas will glow and give off light. Popular types of glow lamps include the NE-2 and the NE-51. The threshold potential is usually somewhere between 40 and 80 volts, although the lamp will maintain its ionized state at a somewhat lower potential, E_H.

An example of a neon glow lamp relaxation oscillator is shown in Fig. 26-1, and the circuit waveforms are shown in Fig. 26-2. Capacitor C is connected in parallel with neon glow lamp I1. A resistor is connected between the lamp-capacitor combination to the power supply and has the effect of limiting the line current. The lamp will assume a low resistance when the gas is in the ionized state, and that means the possibility of a fatal current flowing. Resistor R also figures prominently in the timing of the circuit.

The operation of the neon glow lamp relaxation oscillator is shown in the waveforms of Fig. 26-2. Note that the output waveform, E_0, takes on the form of a semisawtooth after one cycle. At the instant the voltage is applied, the capacitor voltage is zero but will begin rising as the current through the resistor charges capacitor C. The neon glow lamp has a high impedance at this time, so it will draw no current. All of the current through R goes to charge C. The voltage increases according to the well-known expression:

$$E_0 = E(1 - e^{-T/RC}) \qquad \textbf{Equation 26-1}$$

where E_0 is the output voltage, E is the supply voltage, e is the base of the natural logarithms, T is the time after the cycle begins in seconds, R is the resistance in ohms, and C is the capacitance in farads.

The voltage will continue to increase according to Equation 26-1 until it reaches the ionization threshold of the neon lamp. At that time, the resistance between the electrodes inside of the lamp drops to a very low value, causing the charge stored in the capacitor to discharge into the lamp. The voltage waveform therefore drops to the minimum holding voltage; the gas in the lamp will remain ionized to a lower voltage than is required to cause initial ionization. When the capacitor voltage has dropped to holding voltage E_H, the gas inside the lamp deionizes, and the lamp resistance increases to a very high value. The capacitor will start to charge again from E_H to E_t, according to the rate set by Equation 26-1. The output sawtooth will oscillate back and forth between E_H and E_t at a frequency that is a function of supply voltage E; the RC time constant of the network; the ionization threshold of the neon

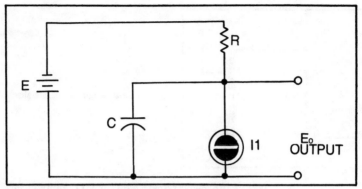

Fig. 26-1. Neon glow lamp relaxation oscillator circuit.

glow lamp and the minimum holding potential of the glow lamp. It is not easy to come up with a simple expression for the frequency of oscillation for this type of relaxation circuit.

UNIJUNCTION TRANSISTOR RELAXATION OSCILLATORS

The unijunction transistor (UJT) shown in Fig. 26-3 has only one pn junction and two bases! Actually, the "two" bases are a single channel, much in the manner of the junction field-effect transistor. If the emitter is reverse biased, no current will flow into the base region from the emitter circuit. But if the emitter junction is forward biased, a very large current will flow in the emitter-base path. Like all pn junctions, the emitter-base junction of the UJT requires a certain minimum forward bias, typically 0.6 to 0.7 volts, before it comes totally forward biased.

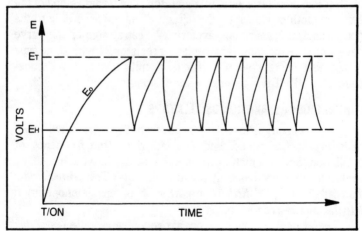

Fig. 26-2. Timing waveform of Fig. 26-1 circuit.

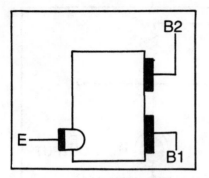

Fig. 26-3. Unijunction transistor.

Figure 26-4 shows the circuit for a simple UJT relaxation oscillator. Timing is a function of the RC time constant of R1/C1. Current I1 will charge capacitor C1 at a rate proportional to the RC time constant and the applied V+ voltage. The waveform across the capacitor will build up in the manner normal for RC circuits until the capacitor voltage exceeds the potential required to forward bias the emitter-base junction of the UJT. When this bias is exceeded, the junction breaks down, and the current stored in the capacitor is dumped through the emitter-base junction. The RC time constant of the capacitor and the emitter-base intrinsic resistance is much less than R1C1, so the discharge slope of the capacitor waveform is much more rapid.

The waveform of this circuit is a semisawtooth (we take the output across the capacitor for a sawtooth, and across R3 for a pulse). In some cases, the UJT relaxation oscillator is linearized to make it into a better sawtooth generator. The leading slope of the semisawtooth produced by this circuit is curved according to the normal capacitor-charging waveform. We can straighten this curve out—linearize the ramp—by replacing resistor R1 with some type of constant current source. These circuits could be a "diode-connected" JFET transistor or a pair of bipolar transistors.

TUNNEL DIODE RELAXATION OSCILLATORS

The relaxation oscillators covered thus far have been resistor-capacitor timed, and this fact limits the frequency of oscillation. Such oscillators will not operate far into the frequency spectrum; they are basically audio oscillators. The *tunnel diode*, also called the *Esaki diode* for its inventor, is capable of operation into the microwave region.

These devices operate on a phenomenon called *negative resistance* or *negative conductance*. If a negative resistance device is

264

connected across a resonant LC tank circuit, it will ring the tank and thereby produce oscillations.

Figure 26-5 shows two popular symbols for the tunnel diode, while Fig. 26-6 shows the *I-versus-E* curve for a tunnel diode. At applied potentials less than some critical value, E_t, the device operates in the normal manner for diodes. The current through the device increases proportional to the applied voltage; (it obeys Ohm's law). This region is called the *positive differential resistance* (PDR) region. At applied potentials greater than E_t, however, the current *decreases* for increasing potentials. This is exactly the opposite of Ohmic behavior—negative resistance! This region of operation is called the *negative differential resistance* (NOR) region.

We can make the forward biased tunnel diode oscillate merely by placing it in parallel with a resonant tank circuit. At microwave frequencies, the diode would be placed at a critical point in a piece of waveguide or cavity. In the circuit of Fig. 26-7, however, a crystal resonator is used to set the operating frequency of the tunnel diode oscillator. The circuit parameters shown in Fig. 26-7 are for the 25 to 30 MHz band. Such a circuit can be used as a micropower CB transmitter, or as a marker for alignment of CB radios.

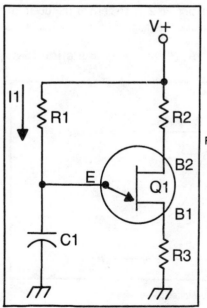

Fig. 26-4. UJT relaxation oscillator.

265

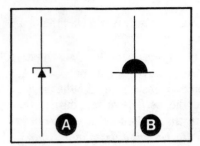

Fig. 26-5. Tunnel diode (A) and alternate circuit symbol (B).

FEEDBACK OSCILLATORS

Figure 26-8 shows the block diagram for a feedback oscillator. The forward block is an amplifier element that has an open-loop (no feedback) gain, A_{vol}. The feedback network may be a simple resistance network or a complex impedance, and its gain is β. Actually, "gain" is usually a loss, so there will be a negative sign. From ordinary feedback theory, we know that the gain of this circuit with feedback A_v is given by the expression:

$$A_v = \frac{A_{vol}}{1 + A_{vol}\beta} \qquad \textbf{Equation 26-2}$$

If this circuit is to oscillate, the closed-loop gain, A_v, must be made infinite. Since A_{vol} will have a finite value (even in an operational amplifier), the only way that we can make the gain infinite is to make the denominator of Equation 26-2 go to zero at the desired frequency of oscillation. To make this happen, then, we need to satisfy the condition

$$|1 + A_{vol}\beta| = 0 \qquad \textbf{Equation 26-3}$$

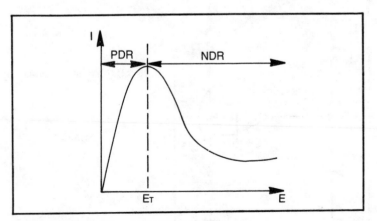

Fig. 26-6. Characteristic negative resistance curve.

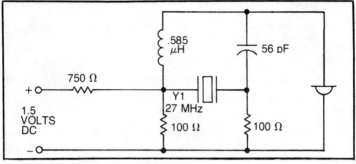

Fig. 26-7. Oscillator circuit using NDR curve.

or $$|A_{vol}\beta| = 0 \qquad \textbf{Equation 26-4}$$

In addition, it is also necessary that the phase shift of the circuit be such that the signal at the input of the amplifier is in phase with the signal at the output. Because most amplifiers used in these applications are inverters, we will need a second 180-degree phase shift in the feedback network. We state our phase shift requirements in the form:

$$\Theta(A_{vol}\beta) = 0 \qquad \textbf{Equation 26-5}$$

These two criteria (Equations 26-4 and 26-5) are known collectively as the Barkhausen criteria. Both must be satisfied for the circuit to oscillate.

Figure 26-9 shows a typical feedback oscillator. Although an operational amplifier is shown, the active element in the forward path could be any form of amplifier that produces a 180-degree phase shift. The feedback factor (β) is given by the expression:

$$\beta = \frac{Z_2}{Z_1 + Z_2} \qquad \textbf{Equation 26-6}$$

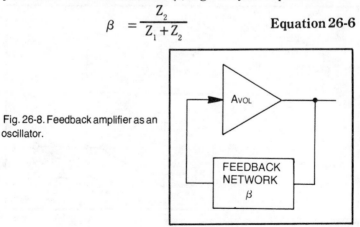

Fig. 26-8. Feedback amplifier as an oscillator.

267

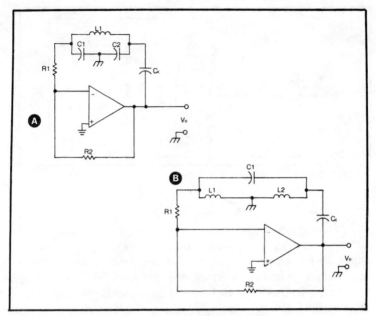

Fig. 26-9. Colpitts oscillator (A) and Hartley oscillator (B).

The impedances shown in this circuit are resistances, but they could be almost any combination of resistance, inductive reactance or capacitive reactance. In general, we can make an oscillator if impedances Z1 and Z2 are of one form and Z3 is of the opposite type. For example, if we use capacitors for Z1 and Z2, an inductor must be used for Z3.

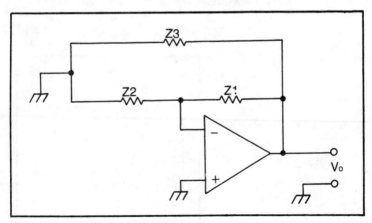

Fig. 26-10. General form of oscillator circuit.

Colpitts Oscillator

Figure 26-10A shows the circuit modified for operation as a *Colpitts oscillator*. We know certain facts about the impedances in this circuit:

$$Z_1 = \omega L_1 \qquad \text{Equation 26-7}$$

$$Z_2 = -1/\omega C_1 \qquad \text{Equation 26-8}$$

$$Z_3 = -1/\omega C_2 \qquad \text{Equation 26-9}$$

We can derive an equation for *resonance*, or the frequency of oscillation, by summing these expression to zero. The result is

$$f = \frac{1}{2\pi \sqrt{L_1 C}} \qquad \text{Equation 26-10}$$

We define C as the total capacitance of the series combination of C1/C2:

$$C = \frac{C1C2}{C1 + C2} \qquad \text{Equation 26-11}$$

The minimum gain of the circuit is the ratio C1/C2, and C2 must be made greater than C1 if the total gain is to be less than unity.

Hartley Oscillator

The classical *Hartley oscillator* is shown in Fig. 26-10B. In this circuit, we have reversed the roles of the impedances; the inductance is split, and the capacitor forms Z3. The impedances of the circuit are:

$$Z1 = \omega L1 \qquad \text{Equation 26-12}$$

$$Z2 = \omega L2 \qquad \text{Equation 26-13}$$

$$Z3 = -1/\omega C1 \qquad \text{Equation 26-14}$$

The frequency of oscillation is:

$$f = \frac{1}{2 \ (L1 + L2)C1} \qquad \text{Equation 26-15}$$

The minimum gain is L2/L1, while the actual gain is R2/R1.

Chapter 27
RC Sine Wave
Oscillator Circuits

The sine wave is the only "pure" AC waveform. All other waveforms are made up of a fundamental sine wave plus a summation of harmonics. These harmonics are described mathematically as a Fourier series of sines and cosines that are integer related and may be offset with a small phase angle from each other. The Fourier series of a pure sine wave contains just one term, however, and that is the frequency of the fundamental itself. In practical terms, the sine wave is the only waveform that will not show any harmonics when examined with a wavemeter or spectrum analyzer.

The sine wave is used for a lot of different purposes, included testing of transmitters and amplifiers. We would use the purest sine wave available if we were trying to measure the total harmonic distortion (THD) of an amplifier. If we used any other waveform, the harmonics present in the driving signal would be added to the apparent distortion of the amplifier, and that simply won't do. In this chapter, we will examine some methods for obtaining a sine wave from an electronic oscillator. Included will be the RC phase shift oscillator, the *Wien-Bridge* oscillator and the twin-T oscillator. We will also discuss a method for obtaining sine waves from squares or triangles and why that is sometimes a better method than using a regular sine wave oscillator.

RC PHASE SHIFT OSCILLATORS

An example of a simple RC phase shift oscillator is shown in Fig. 27-1. The amplifying element in this circuit is a junction

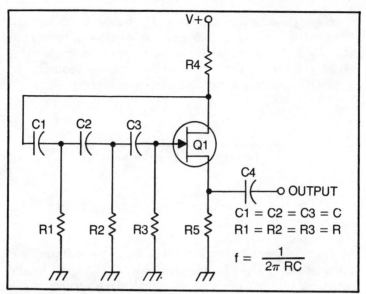

Fig. 27-1. RC phase shift oscillator.

field-effect transistor (JFET), but any of the standard active devices (including operational and other IC amplifier devices) could be used instead of the JFET.

There are two basic criteria for oscillation in any electronic circuit: a feedback signal that is in phase with the input signal and a gain of unity or a little more around the loop. In the circuit of Fig. 27-1, a JFET transistor is connected in the common-source mode, so there is an inherent 180-degree phase shift due to the active element of the RC oscillator. What we need, then, is a feedback network that will provide another 180 degrees (180 + 180 = 360) at the desired frequency of resonance. The required shift is supplied by a three-section RC network (R1/C1, R2/C2 and R3/C3). It is not actually necessary for the three sections to be identical, although that has become the standard practice of circuit designers. It is, then, the usual practice to make the sections each provide a 60-degree phase shift, so the total will be 3 x 60 degrees, or 180 degrees.

In some cases, the resistors are actually potentiometers ganged on the same shaft so that the frequency of the phase shift oscillator can be varied from the front panel of the oscillator cabinet. Different frequency ranges are supplied by making C1 through C3 part of a bank of capacitors that are selected by a range or bandswitch. If we need a fine control over frequency, then one of

271

the resistors (usually R3) can be made from a series combination of a fixed resistor and a potentiometer. That will allow us to vary the phase shift of the last RC section to trim for errors in the operating frequency. Assuming that all three RC sections are identical, the operating frequency is obtained from the expression:

$$f = \frac{1}{2\pi RC \sqrt{6}} = \frac{1}{2\pi (2.45)RC} \qquad \textbf{Equation 27-1}$$

where f is the frequency in hertz, R is the resistance in ohms, and C is the capacitance in farads.

☐ **Example:**

Find the operating frequency of an RC phase shift oscillator such as shown in Fig. 27-1 if the resistance is 22K ohms and the capacitance is $0.002\ \mu F$. Assume that all three sections of the feedback network are equal.

$$f = 1/(6.28)(2.2 \times 10^4 \text{ ohms})(2 \times 10^{-9} \text{ farads})(2.45)$$

$$= 1/(2.8 \times 10^{-4})(2.45)$$

$$= 1458 \text{ hertz}$$

It is the usual practice to rewrite Equation 27-1 so that we solve for the resistance and *select* the frequency and capacitance. We would know the required frequency from the application, and it is easier to select capacitor values than resistor values, because capacitors have fewer readily available values than resistors. The capacitor value could be selected according to gut feeling, or from total ignorance. Then a trial value for the resistor would be calculated and checked to see if it were reasonable.

If we perform a circuit analysis on the phase shift oscillator, we would find that the beta factor of the feedback loop is approximately 1/29. This means that the amplifier element must have a gain of a least 29 for the circuit to oscillate. In actual practice, it is advisable to make the amplifier gain somewhat larger than needed—about 10 percent larger. This will tend to keep the oscillator running even after some of the components have changed value and if the gain of the amplifier element is somewhat less than anticipated. Some types of amplifier will lose gain over a period of time.

WIEN-BRIDGE OSCILLATOR

The Wien-bridge is an RC bridge similar in form to the famous Wheatstone bridge. Two of the arms of the Wien bridge are resistances, while the other two are RC networks. One of the RC networks is a series circuit, while the other is a parallel circuit. An example of a Wien bridge oscillator is shown in Fig. 27-2. We know that voltages E1 and E_0 are always in phase with each other, and both are 180 degrees out of phase with amplifier input voltage E2. The feedback loop is degenerative—and therefore stable—at all frequencies other than the resonant frequency, which is given by the expression:

$$f_{Hz} = \frac{1}{2\pi\sqrt{R_3 R_4 C_1 C_2}}$$ Equation 27-2

or if R3 = R4 and C1 = C2,

$$f_{Hz} = \frac{1}{2\pi\sqrt{R^2 C^2}} = \frac{1}{2\pi RC}$$ Equation 27-3

Voltage E3 is out of phase with E_0, but at frequency F, described by Equation 27-2, voltages E_0 and E_3 are in phase with each other; hence, the phase angle is 360 degrees. Since E3 is applied to the noninverting input of the amplifier, this circuit will oscillate.

The output amplitude of the Wien-bridge oscillator may tend to be a little unstable if only RC elements are used in the device. To overcome this instability, a lamp (I1 in Fig. 27-2) is placed in one arm of one of the resistances. The lamp stabilizes the output amplitude and prevents the amplifier from saturating, because of its nonlinear voltage-current characteristics. In most designs, lamp I1 is operated at a current level just below incandescence.

TWIN-T OSCILLATOR

There are a variety of circuits using just resistors and capacitors that will permit an oscillator to work. All that is needed is a RC phase shift network that provides the needed phase shift (either 180 or 360 degrees, depending upon the design). One possible network is the twin-T RC circuit shown in Fig. 27-3. The same is derived from the fact that there are two parallel-connected T-networks between the output and input of the amplifier element. One of the T networks consists of two series resistors and a shunt capacitor, while the other is exactly the opposite: two capacitors and a resistor. The operating frequency is given by the expression:

$$f = \frac{1}{2}\pi RC 8 = 1/17.8RC$$ Equation 27-4

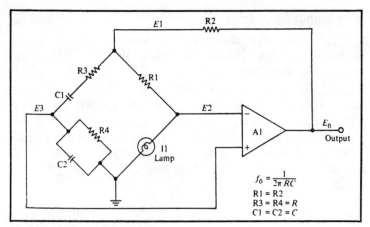

Fig. 27-2. Wien-bridge oscillator.

There are two additional variations on the twin-T oscillator design, and these are called *bridged-T oscillators* (Fig. 27-4). The circuit shown in Fig. 27-4A is a resistive version (resistance R is shunted across a single-T network), while that in Fig. 27-4B is a capacitive bridged-T oscillator.

GENERATING SINE WAVES FROM A SQUARE WAVE OR TRIANGLE WAVE

The sine wave can be generated from *any* other form of wave if the harmonics of the fundamental are filtered out. Recall that the square and triangle waveforms are made from a fundamental sine wave, plus various combinations of its harmonics. The different waveshapes obtained from any given fundamental depend upon which harmonics are present, their relative amplitudes and their

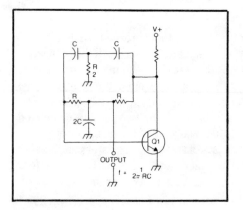

Fig. 27-3. Twin-T oscillator.

274

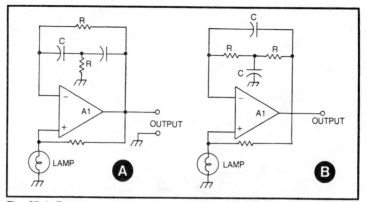

Fig. 27-4. Resistive version of a bridged-T oscillator at A, and a capacitive version of a bridged-T oscillator at B.

relative phase angles with the fundamental and each other. If all of the harmonics above the fundamental are filtered out, the only remaining signal will be a sine wave at the fundamental frequency. The purity of the sine wave can be quite good, but it is dependent upon the ability of the filtering process to cope with the harmonics.

It is a general design trend to pass the square wave through a series of low-pass filters that have a sharp attenuation slope above frequency f_0. The signal can then be passed through a narrow bandpass AC amplifier that will only accept the fundamental frequency. Some of the best sine wave generators on the market use this approach.

Just why would anyone want to make a sine wave by filtering a square wave? After all, we have some perfectly wonderful sine wave oscillators, don't we? Those circuits are not all that wonderful in some respects. For one thing, the sine wave oscillators are a little difficult to make variable over a wide range. The RC phase shift, Wien-bridge and twin-T oscillators all require the switching of multiple resistors and capacitors in order to make either frequency of frequency range changes. Many square wave circuits, on the other hand, will accommodate frequency changes by varying one potentiometer and range changes by changing one capacitor. In addition, the square wave and triangle wave circuits are usually far more able to sustain a stable output voltage than are some of the sine wave circuits. We could make the square wave circuit a lot more stable and find it a lot easier to shift frequencies. Then we could use one of the variable active filter circuits in order to reduce the amplitude of the harmonics without affecting the fundamental.

275

Chapter 28
LC Oscillator Circuits

Oscillators that operate in the rf region very frequently use either a *piezoelectric crystal resonator,* or an *LC tank circuit.* The crystal oscillator is discussed at length in Chapter 31, so we will cover only the LC oscillator in this chapter.

LC oscillators use inductance (L) and capacitance (C) elements in a resonant tank circuit to achieve oscillation. What is this process called *resonance*? The impedance of an LCR network is a complex value that must take into account such factors as resistance, inductive reactance and capacitive reactance. The two reactances offer a phase shift to alternating currents; the instantaneous relationship between current and voltage will change if the circuit is reactive. The current and voltage will remain in phase with each other if the circuit impedance is purely resistive, but an inductor will oppose changes in the *current* flowing, while not affecting the voltage. If a sine wave is applied to an inductor, the current will lag behind the voltage; in a pure inductance, the phase shift is 90 degrees. A capacitor is exactly the opposite: It opposes changes in voltage while not opposing the flow of current. This means that the voltage will lag behind the current. Again, a perfect capacitive reactance will produce a -90-degree phase shift.

The impedance in an LCR tank circuit is defined as the vector sum of the components produced by the resistance, inductance and capacitance:

$$Z = \sqrt{R^2 + (X_L - X_C)^2} \qquad \textbf{Equation 28-1}$$

where Z is the impedance in ohms, R is the resistance in ohms, X_L is the inductive reactance in ohms, and X_C is the capacitive reactance in ohms.

The inductive and capacitive reactances are defined in terms of frequency:

$$X_L = 2\pi f L \qquad \textbf{Equation 28-2}$$

$$X_C = \frac{1}{2\pi f C} \qquad \textbf{Equation 28-3}$$

where f is the frequency in hertz, L is the inductance in henrys, C is the capacitance in farads, and X_C and X_L are as defined earlier.

Resonance is a condition in which the inductive and capacitive reactances are equal in magnitude, or $X_L = X_C$. Since these are vector quantities that have opposite signs, at resonant frequency F, the inductive and capacitive reactances cancel each other, leaving only the circuit resistance. See Equation 28-1 and determine what happens when $X_L = X_C$.

SHOCK EXCITING AN LC CIRCUIT

An LC circuit is shown in Fig. 28-1. We will assume that there is no resistance in the circuit and that the only resistive losses are the small wire winding resistances in the inductor. Switch S1 is normally open, so no current will flow from the battery into the LC tank circuit.

Now let's consider what happens if switch S1 were momentarily depressed, allowing current to flow into the tank circuit. The tank current will immediately charge capacitor C, but as soon as the switch opens again, the charge in the electrostatic field of the capacitor is dumped back into the circuit and flows through the inductor. The current in the inductor forms a magnetic field around the inductor. When this field collapses an instant later, it generates a countercurrent that once again charges the capacitor. This exchange of energy takes place back and forth between the electrostatic field of the capacitor and the magnetic field of the inductor, producing a *sinusoidal oscillation* at the resonant frequency of the tank circuit. If the tank elements were perfect, the current—once induced—would flow back and forth forever. Several experiments have been performed in which supercooled tank

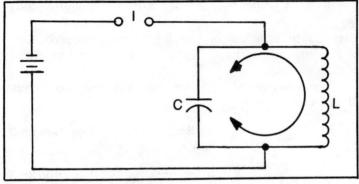

Fig. 28-1. Shock-excited tank circuit.

circuits have been kept at − 250° F. The experimenters were able to keep the oscillatory current going for a long period of time. But in real LC tank circuits, operated at room temperatures, there are substantial losses. The series DC resistance of the inductor winding, for example, is a considerable loss. If any external device is coupled to the LC tank circuit, it will draw energy from the tank and further increase the losses. The sum of the losses will conspire to reduce the amplitude of the oscillations on each successive cycle, as shown in Fig. 28-2. The amplitude of the oscillation will die out in an exponential decay manner, as shown.

Notice that the general waveshape of the oscillatory signal is *sinusoidal*, yet the excitatory current was a step-function pulse (the switch closure). This difference brings us to the matter of the *flywheel effect*. The sinusoidal current being produced by a pulse input is the flywheel effect, and it draws its name from analogy to the action of a mechanical flywheel: One push will set it rotating for several cycles.

We can use the LC shock excitation phenomenon to make oscillator circuits. If either the capacitance or the inductance in the LC tank circuit is variable, we will make a *variable frequency oscillator*, or *VFO*. The tuners of most classical radio receivers are VFOs in which the capacitance is made variable. Most car radios, however, plus a few consumer table model radios made during the copper shortage days of World War II, use a variable inductor as the tuner. In those radios, a permeability core is moved in and out of the inductor form to increase or decrease the inductance of the coil.

LC tank circuits can be used in several different types of oscillator circuits, of which several will be considered here: Armstrong, TGTP, Hartley, Colpitts, and so forth. We will also

278

consider a version of the Colpitts oscillator, called the Clapp oscillator. Incidentally, each of these oscillator types can be identified by one or more salient features, and it is these features that can get you past a school, trade or FCC license examination.

Almost every oscillator circuit can be built in any of several different configurations. The two classes that are most common can be divided into series feed and shunt (parallel) feed. A *series feed oscillator* is one in which the collector (or anode or drain) current passes through the inductor of the LC tank circuit. In the *shunt feed circuit,* the LC tank circuit is cold to DC; in other words, none of the DC current for the collector (or drain or anode) flows through the inductor of the tank.

ARMSTRONG OSCILLATORS

The Armstrong oscillator was one of the first types known and is named for its inventor, Major Edwin Armstrong. In most classical versions of this circuit, the active element was a vacuum tube, but in the modern example of Fig. 28-3, a field-effect transistor is used instead. Any active electron device, however, could be used, including bipolar transistors and integrated circuits.

An LC tank circuit, L1/C1, is connected between the gate of the JFET and ground. Capacitor C2 serves as a DC blocking capacitor, and—in vacuum tubes—it would form a little grid leak bias. The parallel resonant tank circuit shunts to ground all frequencies except its resonant frequency.

Feedback from the drain of the JFET is provided by a link-coupled *tickler coil* (L2). When power is first applied to the JFET, the drain current is rising from zero to its normal quiescent value. This changing current creates a changing magnetic field around L2. Because L2 is normally closely coupled with coil L1,

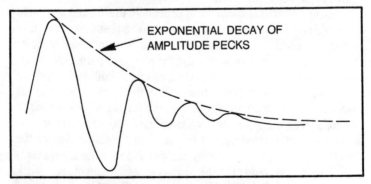

EXPONENTIAL DECAY OF AMPLITUDE PECKS

Fig. 28-2. Exponential decaying oscillatory waveform.

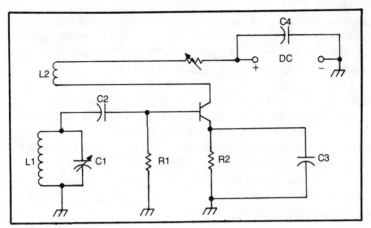

Fig. 28-3. Armstrong oscillator.

this magnetic field will shock excite the L1/C1 tank circuit. The oscillations developed across the shock-excited tank circuit become an input signal for the JFET; it is impressed across the gate-source terminals of the JFET. The oscillatory voltage is amplified by the JFET, creating a drain current variation that is identical in shape to the gate signal. This signal, in turn, sets up variations in the magnetic field of the tickler coil, which are induced into the L1/C1 tank circuit. Once this action begins in earnest, the oscillations become sustained, and the tank will be re-excited once every AC cycle. The frequency of oscillation will be

$$f = \frac{1}{2\pi\sqrt{LC}}$$ **Equation 28-4**

where f is the frequency in hertz, L is the inductance in henrys, and C is the capacitance in farads.

Some Armstrong oscillators, especially those used as a radio detector, contain a *regeneration control* that adjusts the amount of feedback needed to sustain the oscillations. In some older models, the regeneration control was in the form of having the tickler coil position variable with respect to the tank circuit coil. When the two coils are positioned such that their respective axes are 90 degrees from each other, there is practically zero feedback. But when the two coils share the same axis, the coupling is close to 100 percent, so the feedback is maximum. It was the usual practice to place the tickler coil on a gimbal mounting inside L1. In bipolar transistors, the regeneration could take the form of an adjustable bias network for the transistor. In the JFET case shown in Fig. 28-3, however, a series potentiometer (R3) is used to adjust the current flowing in the drain circuit of the JFET. The Armstrong oscillator is not very popular today, but it is seen occasionally as the local oscillator (LO) or beat frequency oscillator (BFO) in certain types of low-cost radio receivers.

TGTP OSCILLATORS

The TGTP oscillator receives its name from *tuned-grid–tuned-plate,* a term that is now obsolete by the eclipse of the vacuum tube as an amplifying element. This term, however, persists even today when solid-state elements are likely to be used in the "TGTP" oscillator. We will continue to use the TGTP designation, even when we *really* mean *tuned-collector–tuned-base,* or *tuned-drain–tuned-gate.* Remember: TGTP = TCTB = TDTG, except for the oscillating active element.

The oscillator (Fig. 28-4) works because of the interelement, or interelectrode, capacitances. All tube elements, or transistor junctions, possess some amount of capacitance. In an rf amplifier circuit, these must be cancelled in the process called neutralization. But in the TGTP oscillator, we take advantage of these capacitances. Capacitances C_{cb} (collector-base) and C_{be} (base-emitter) form a capacitive voltage divider inside the transistor. The collector-emitter current, in effect, creates an in-phase voltage drop across the base-emitter capacitance. This forms a valid input signal for the transistor and causes it to be amplified. After a few cycles, this oscillation is sustained. The signals are selected by the collector and base LC tank circuits. Only those signals at the resonant frequency of the tanks, which are shock-excited by the transistor currents, will oscillate.

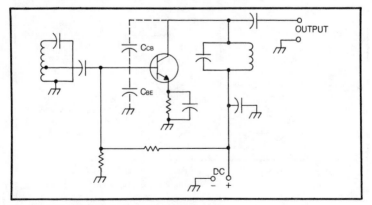

Fig. 28-4. Tuned-based—tuned-collector oscillator.

The collector and base tank circuits are tuned to approximately the same frequency. They are at substantially the same frequency, but are usually detuned from each other by a small amount in order to make the oscillation more stable. When they are identically tuned to the same frequency, the oscillation will tend toward instability.

In vacuum-tube circuits, the interelectrode capacitances are usually quite sufficient to maintain oscillation. In fact, it is usually *so* sufficient, that rf amplifier designers who tune both grid and plate have fits trying to neutralize the circuit. But in transistor circuits, the interelement capacitances are usually somewhat smaller. In those cases, there might be an external capacitor connected either from collector to base, or collector to emitter.

HARTLEY OSCILLATORS

Figure 28-5 shows the Hartley oscillator circuit. The principal feature of this type of oscillator is the *tapped coil* in the LC tank circuit as part of the feedback network. Don't confuse the tapped coil used in some rf amplifiers as a means for impedance matching with the tapped coil of the Hartley oscillator. If the coil is part of a feedback network, it is an oscillator and *not* an impedance matched amplifier.

In the Hartley oscillator circuit shown in Fig. 28-5, the emitter-collector current flows in part of the coil (L1b). When the oscillator is first turned on, the rising emitter current creates a varying magnetic field around the portion of L1 in which it flows.

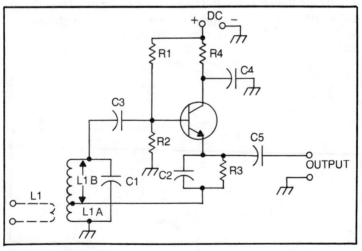

Fig. 28-5. Hartley oscillator.

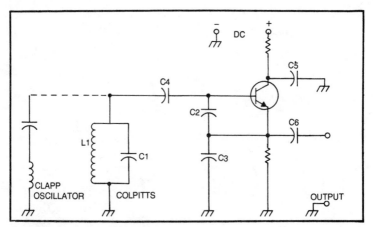

Fig. 28-6. Colpitts and Clapp oscillators.

This induces a current into the rest of L1 by autotransformer action, thereby shock exciting the LC tank. The oscillatory current in the tank presents a voltage to the base of the transistor, and this is seen as a valid sine wave input signal. When this signal is amplified, it creates a varying emitter-collector current identical in waveshape to the base voltage (a sinusoid). This further shock excites the LC tank via L1a, and the oscillation becomes sustained.

COLPITTS OSCILLATORS

The Colpitts oscillator is similar to the Hartley in that it uses a tapped reactance to provide the feedback required to begin and sustain oscillation of the active element. But in the Colpitts—and its relative, the Clapp oscillator—the feedback is a *capacitive voltage divider*, instead of the inductive divider used in the Hartley. An example of a Colpitts oscillator is shown in Fig. 28-6. The feedback voltage divider (C2/C3) forms the feedback network for this circuit. If the LC tuned circuit is a parallel resonant tank, the oscillator is a Colpitts oscillator. If the tuned circuit is a series resonant tank, it is called a Clapp oscillator.

The transistor emitter current at start-up begins at zero, so the voltage drop across emitter resistor R1 is zero. After the instant of turn-on, however, the voltage across R1 begins to climb, impressing a changing voltage across voltage divider C2/C3. This signal is seen by the transistor as a valid input signal and is amplified, causing an oscillatory shock to the LC tank circuit.

The Colpitts and Clapp oscillators are generally considered best for high-frequency (into the low VHF) work. The earlier Hartley oscillator is useful for lower frequency oscillators.

Chapter 29
RC Monostable Multivibrators

The monostable multivibrator is a circuit that will produce a single output pulse of constant duration and amplitude every time a trigger pulse is received at the input. For this reason, it is also called *one-shot* circuit. So what use is a circuit that outputs a single pulse, for each input pulse, on a one-for-one basis? Actually, there are several uses for the one-shot stage. It will, for example, make a dandy pulse delay line. If two one-shots are connected in series, the first can provide the delay, and the second can reproduce the input pulse. Suppose we have a 100-microsecond pulse and want to delay it 1.25 milliseconds. We could set the duration of the first one-shot at the desired delay of 1.25 ms. The second one-shot would then be programmed to 100 μS, and it would reproduce the original pulse. By making the first one-shot such that it has a variable time delay, we can also vary the repetition rate of the pulse train.

Another use of the one-shot is to clean up a pulse. We sometimes have poor pulses in circuits, especially after transmission over a coaxial cable or through some other medium. The one-shot will reproduce each input pulse with the nice sharp features needed by most electronic circuits.

One-shot circuits are also used to synchronize some digital electronic circuits, or to form bounceless pushbuttons. Many digital circuits will change inappropriately in response to the bounce noise present on most switches. The one-shot will output a single pulse that the digital circuit sees as a switch closure and that remains HIGH for the duration of the bounce noise of the switch. As a result, the digital circuitry sees only one pulse (as it should) even though a few dozen may actually have occurred.

Analog tachometer circuits use one-shot circuits to good advantage. This type of electronic device, which may be as simple as an automotive tach, or a supposedly sophisticated analog heart rate/respiration rate meter from some medical electronics device, produces a DC voltage that is proportional to the frequency of the input signal. This DC voltage is then displayed on a meter that is calibrated in units of revolutions per minute, beats per minute, breaths per minute and so on.

TRANSISTOR ONE-SHOTS

The earliest semiconductor devices used to make monostable multivibrators were the transistors (pnp and npn). These circuits were relatively elementary and could be made with only two similar transistors. An example of such circuit is shown in Fig. 29-1, while the timing diagram is shown in Fig. 29-2. The circuit consists of two cross-coupled npn transistors. The circuit is said to be cross-coupled because the collector of each transistor is connected to the base of the other.

A monostable multivibrator receives its name from the fact that it will have only *one stable state*. A trigger pulse will cause the circuit to switch to the unstable state, but it will remain there only for a certain specified period of time. The circuit will then revert automatically to the stable condition. In the circuit of Fig. 29-1, the stable state will see transistor Q1 turned off by a high negative bias applied to its base from the V− power supply. Transistor Q2, however, will be turned on to the point of saturation.

In the stable state, transistor Q1 is turned off. Its collector voltage therefore rises to V+, charging capacitor C1. The voltage across capacitor C1 will be approximately $(V+) - (V_{beq2})$.

The circuit will remain dormant, with V_o very low (equal to the $V_{ce(sat)}$ of transistor Q2) until a trigger pulse is received. We can trigger this circuit either by turning on transistor Q1 or by turning off Q2. In this circuit of Fig. 29-2, we have elected to trigger the one-shot by turning off transistor Q2. The trigger pulse is capacitor-coupled to the base of Q2 through isolating diode D1. This pulse (see Fig. 29-2) is applied to the base of transistor Q2, causing it to turn off. The voltage on the base of transistor Q2 is normally a small positive value V_{be}. But when the trigger pulse is received, transistor Q2 will momentarily turn off, causing its

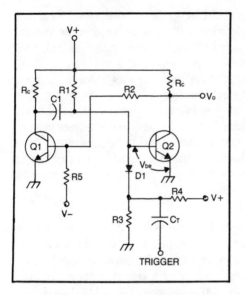

Fig. 29-1. Transistor monostable multivibrator.

collector voltage to rise to V+. When this occurs, the contribution of V+ towards the bias of transistor Q1 through resistor R2 increases sufficiently to drive Q1 into saturation. This will ground the Q1 end of capacitor C1, causing the base of Q1 to see a high negative voltage. The voltage across C1 will keep Q2 turned off. During this period, the collector voltage of transistor Q1 has dropped to $V_{ce(sat)}$, while that of Q2 has risen to approximately V+.

The capacitor voltage does not remain constant, however. The capacitor is discharged by the current flowing in resistor R1. At a time, $T = 0.69 R_1 C_1$, the situation reverses, and the circuit goes back to its stable state. Transistor Q2 will no longer be biased off by the voltage across capacitor C1. When transistor Q2 turns back on, the Q2 collector voltage drops back to $V_{ce(sat)}$, and this removes the bias on the base of Q1. The output can be taken from either collector, depending upon whether a normally-HIGH or normally-LOW output state is desired.

We cannot, however, retrigger this circuit for a certain period of time after the one-shot returns to its stable state. This is the time period required for C1 to recharge to $(V+) - V_{be}$, and it is governed by its own value and that of resistor R_c. This period of time is called the *refractory period*.

The use of bipolar transistors in one-shot design has more or less been superseded by the use of integrated circuitry. Some ICs are special-purpose monostable multivibrator chips, some are

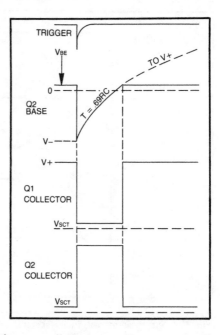

Fig. 29-2. Timing waveform
of Fig. 29-1 circuit.

timer chips, while still others are little more than operational
amplifiers cleverly configured.

OPERATIONAL AMPLIFIER ONE-SHOTS

An operational amplifier can be made to behave as a one-shot
multivibrator if it is operated in its comparator mode. A *comparator*
is considered to be an amplifier with far too much gain. In the
operational amplifier version, this means using the op amp with no
feedback resistor; the circuit gain is essentially the open-loop gain

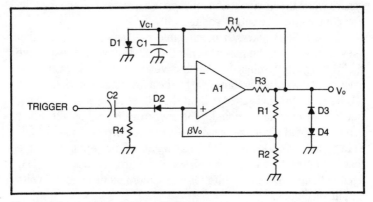

Fig. 29-3. Op amp monostable multivibrator.

287

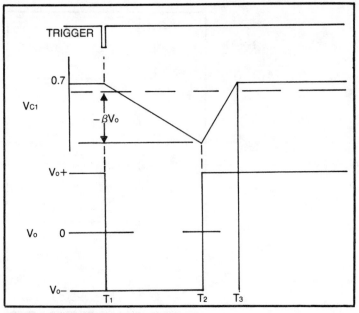

Fig. 29-4. Timing diagram of Fig. 29-3 circuit.

of the device. The basic comparator will issue only three possible output levels. When the voltages applied to the two inputs are equal, the output voltage is zero. When the two input voltages are not equal, however, the output voltage will be either V− or V+, depending upon which input voltage is higher. If the voltage applied to the inverting input is greater than the voltage applied to the noninverting input, the amplifier sees this as a negative differential input voltage, making the output voltage V−. The exact opposite occurs when the noninverting input sees a greater voltage than the inverting input: The output voltage will be V+.

Figure 29-3 shows a monostable multivibrator circuit that uses an operational amplifier, while Fig. 29-4 shows the respective timing diagram. Ordinarily, the noninverting input is kept at some fraction of the output voltage (V_o) by the action of feedback network R1/R2 (resistor R3 is used to limit the output current and need not figure into our calculations). The inverting input, on the other hand, is kept at a voltage equal to the voltage across capacitor C1. In the stable, or dormant, state, this input will be kept at a voltage that is limited to the forward bias voltage drop across diode D1. This voltage will be 0.3 volts for germanium diodes and 0.7 volts for silicon diodes.

Refer to Fig. 29-4. At time T_1, a trigger pulse is received at the noninverting input. Note again the use of a differentiating network (C2/R4) and an isolation diode (D2) in the trigger circuit. The action of the trigger pulse is to momentarily bring the noninverting input to a potential less than the inverting input, causing the operational amplifier output to switch from V_{o+} to V_{o-}. Now that V_o is negative, the negative potential will begin to discharge the small positive charge on capacitor C1 and then cause it to begin charging in the negative direction. During this time period the output voltage will remain at V_{o-} until V_{C1} reaches a potential that is equal in magnitude to the potential applied to the noninverting input. This is designated as $-\beta V_o$ and is defined as:

$$V_{C1} = \left[\frac{R2}{R1 + R2}\right] V_o$$

When the capacitor voltage causes the relationship between the inverting and noninverting inputs to reverse, the output will return to the static condition V_{o+}. The period of this multivibrator is:

$$T = R_1 C_1 \ln \left[\frac{1 + (V_{C1}/V_o)}{1 - \beta}\right]$$

The purpose of zener diodes D3 and D4 is to limit the output voltage to a value ($V_z + 0.7$). The output voltage will always be 0.7 volts higher than the zener potential because of the forward bias potential of the other diode in the circuit. If we want the output waveform to be symmetrical, we would make the zener potentials of both diodes equal. If we need asymmetrical output, on the other hand, we would make the zener potentials unequal, as needed.

Neither of the monostable circuits described thus far in this chapter has been retriggerable. Once a trigger pulse is received at the input, the circuit cannot be retriggered again until the period has timed out and the refractory period has expired. A retriggerable monostable, on the other hand, would respond any time that a trigger pulse is applied. If the trigger pulse were applied before the original period had timed out, the output would remain in the unstable state for one more period. In most cases, designing a retriggerable monostable multivibrator means that we must be able to either dump the charge from the capacitor, or recharge it—depending upon the design—using the trigger pulse.

An example of a retriggerable monostable multivibrator is shown in Fig. 29-5, while the timing diagram is shown in Fig. 29-6. In this circuit, the noninverting input of the operational amplifier is kept at a potential set by resistor voltage divider R1/R2. The

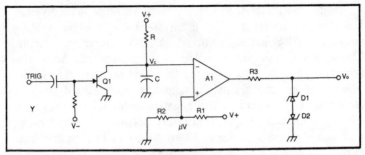

Fig. 29-5. Comparator monostable multivibrator.

inverting input is kept at the capacitor voltage, which is determined by time constant RC.

The heart of this circuit is transistor switch Q1 that is used to dump the charge in capacitor C. This transistor is normally kept biased off, so it will present a high drain-source resistance. This situation allows capacitor C to be charged by the current flowing from V+ through resistor R. But when a trigger pulse is received, transistor Q1 will momentarily turn on and will cause the charge in capacitor C to be dumped. The output voltage will snap HIGH and will remain HIGH until the capacitor has charged to at least the same voltage as on the noninverting input. The capacitor cannot begin charging until the trigger pulse vanishes.

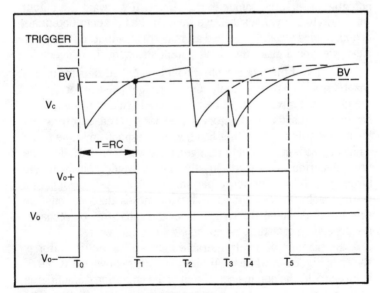

Fig. 29-6. Timing waveform of Fig. 29-5 circuit.

The action of the one-shot over one time period, without retriggering, is shown as period T_0 to T_1 in Fig. 29-6. But look what happens when a second trigger pulse is received before the time period has expired. The circuit is triggered at time T_2, which would normally mean that the period would expire at time $\hat{T}_4$. But at time T_3, a second trigger pulse is received, causing the output period to be extended by one whole cycle. The duration of the output cycle is given by:

$$T = RC \ln(1 + (R2/R1))$$

CMOS ONE-SHOTS

Certain CMOS integrated circuits can be used to make monostable multivibrators. Some of these are special-purpose CMOS chips designed originally as one-shots. Other circuits use ordinary CMOS gates or inverters to form the monostable function.

The simplest type of circuit is not actually a full monostable stable, but it could be called a semimonostable, or half monostable. This circuit (Fig. 29-7) can be built with any CMOS inverter, buffer, or any of the gates (NAND/NOR) that can be made to act as an inverter. In the gate case, we sometimes find that tying the two inputs of a NAND gate or NOR gate together will result in inverter action. The half monostable of Fig. 29-7 will produce an output period that is equal to approximately:

$$T = 0.8\,RC$$

The particular version of the half monostable shown in Fig. 29-7 requires a trigger pulse that snaps HIGH from a zero volts condition. We can obtain the opposite triggering by the simple expedient of connecting the end of the resistor (R) to V+, instead of to ground. The range of periods available with readily obtainable component values is from less than 1 microsecond to over 1 second.

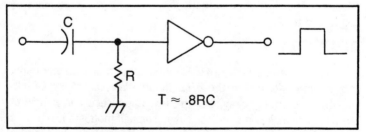

Fig. 29-7. Half monostable circuit.

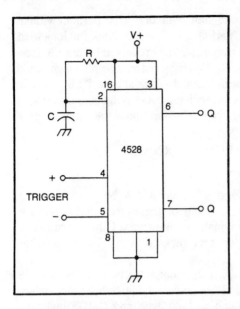

Fig. 29-8. 4528 CMOS
monostable multivibrator.

Figure 29-8 shows the use of a CMOS 4528 device. This CMOS integrated circuit is a dual monostable multivibrator (only one section shown in Fig. 29-8). The period is set by an RC time constant. The timing resistance must be 10K ohms to 10 megohms. The timing capacitor can be anything greater than 20 pF. If a 5 VDC power supply is used, the minimum obtainable pulse duration is approximately 550 nanoseconds, which reduces to 350 nanoseconds if the supply voltage is 10 volts or greater. This circuit uses complementary outputs, meaning that Q and not-Q outputs are available.

Note that there are two trigger inputs. These can be used to determine the polarity of the trigger pulse that will cause the output to change. The + trigger input accepts a positive-going trigger pulse, while the − trigger input accepts a negative-going trigger pulse. In both cases, the unused trigger input must be connected to V+.

TTL MONOSTABLES

There is no easy way to make a one-shot multivibrator using only simple gates from the TTL family. We can make CMOS one-shots relatively easy, but the low impedance and nature of TTL devices makes it a lot less easy. The half monostable circuits are possible but are a little difficult to actually make work. It is a lot easier to write them up in a book or magazine article than it is to actually make them work on the bench. In contrast, the CMOS version is almost trivially easy to make work.

When we want a TTL one-shot, it is sometimes necessary to use one of the special-function integrated circuits on the market, such as the 72121, 74122, and 74123 devices. Unfortunately, these can be a little "flakey" when there are noisy power supplies, improper circuit layout or certain other defects. When periods longer than 1 μS, are desired, it is sometimes better to use another technique or device, such as the 555 to be described next. But for really short duration pulses, these TTL devices cannot be beat. Figures 29-9 and 29-10 show the 74121 and 74122 devices. Each of these is controlled by an RC time constant network and will produce an output period of approximately T = 0.7RC. Values for R range from 2K ohms to 30K ohms (40K ohms in LS-series TTL devices), while the capacitor value may be 10 pF to 10 μF. The trigger signal in both cases must be negative-going from a +5 VDC level. The principal difference between these two devices is that the 74122 is retriggerable, while the 74121 is not.

Note that these TTL devices are high-frequency integrated circuits and must be treated as such in the circuit. They are capable of producing some very short pulses, and this implies high-frequency operation. It is essential that the triggering be proper and that proper circuit layout be observed. Each device should be bypassed right at the power-supply terminals with a disc capacitor in the 0.001 μF to 0.01 μF range.

555 TIMER AS A ONE-SHOT

One magazine article claimed that the 555 timer was one of the all-time most popular integrated circuits and ranks second only to the 741 operational amplifier. This device is an RC timer in an 9-pin miniDIP package that is capable of producing relatively stable output pulses in the 0.1-microsecond to 10-second range. It is a generally well behaved device (there are a couple of exceptions) and is relatively easy to apply, even for a novice. Figure 29-11 shows the basic circuit for the 555 device when it is used as a monostable multivibrator. The output duration is set by the RC combination and is approximately:

$$T = 1.1\,RC$$

The output signal is available at pin No. 3 and will snap HIGH when active, dropping back LOW again when the period of the time

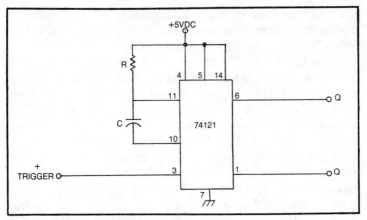

Fig. 29-9. 74121 monostable multivibrator.

constant has expired. The trigger terminal requires a relatively unusual signal. Although it is negative-going it must be able to drop from V+ down to a level of (V+)/3 or less. Some other devices require the voltage to drop almost to zero and will mistrigger if the trigger pulse is sloppy. We cannot really tell much about how this circuit works from the circuit shown, so it may prove helpful to study the internal workings of the 555, at least in block diagram form.

Figure 29-12 shows the internal block diagram of the 555 integrated circuit. This low-cost device is usually supplied in an 8-pin miniDIP package, although several manufacturers also offer devices that are essentially dual (556) 555s and quad 55s.

One reason for the wide popularity of the 555 is its extreme flexibility. It is neither TTL nor CMOS, but it is compatible with both under the right set of circumstances. The 555 is of bipolar

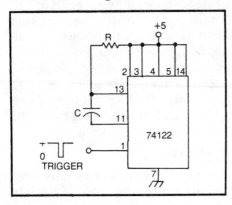

Fig. 29-10. 74122 circuit.

294

construction and can operate with supply voltages that range from 4.5 to 15 VDC. Most authorities seem to indicate a supply voltage between 9 and 12 volts is most optimum. When the 555 output is in the LOW condition, the output terminal will *sink* up to 200 milliamperes of current. When in the HIGH state it will *source* 200 milliamperes.

The range of possible applications for the 555 seems limited only by the imagination of the designer. The 555 is immune (well, almost) to changes in supply voltage; which is something that not every RC oscillator or one-shot can claim. The 555 monostable mode can be used as a timer, delay line, etc. Your own understanding of the internal operation of the 555 will allow you to make much better use of the chip.

The internal block diagram of the 555 is shown in Fig. 29-12. The principal internal stages include two voltage comparators, an RS flip-flop, output inverting amplifier and a discharge transistor (Q1).

A *voltage comparator* is basically an operational amplifier with an open feedback loop. This means that the gain of the stage is essentially the open-loop gain of the operational amplifier. This gain can be as low as 10,000 or as high as 1,000,000. Obviously, only a couple of millivolts difference between the potentials

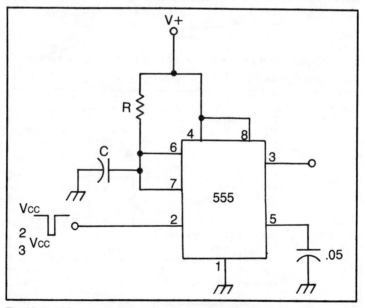

Fig. 29-11. 555 monostable multivibrator circuit for one-shot.

applied to the inputs will cause the output to saturate. If the input voltages are equal, the output voltage is zero. If the input voltages differ by only a tiny amount, however, the output will be saturated either positive or negative.

An *RS flip-flop* is a digital integrated circuit that will SET (make the Q output HIGH and the not−Q output LOW) when a pulse is applied to the *S* terminal, and RESET (make the Q terminal LOW and the not-Q HIGH) when a pulse is applied to the *R* terminal. The RS flip-flop is a bistable circuit. This means that it has two stable states and is equally able to remain in either. It will not change state unless an appropriate pulse is applied to one of the inputs. In the 555, the inputs to the RS flip-flop are controlled by the outputs of the comparators.

Under the initial conditions, at time T_o, the not-Q terminal of the RS flip-flop is HIGH, and this biases transistor Q1 hard on, placing IC pin No. 7 effectively at ground potential. This keeps capacitor C1 discharged, Also, amplifier U1 is an inverter, so the output terminal at pin No. 3 is initially in the LOW state.

Resistors R_a, R_b and R_c are inside the IC and are of equal value (nominally 5000 ohms). These form a *voltage divider* that is used to control the voltage comparators. The inverting (−) input of comparator No. 1 is biased to a potential of:

$$E_1 = \frac{R_b + R_c}{R_a + R_b + R_c} \times (V_{cc})$$

Because all three resistances are the same, we can infer from this equation that the output of comparator No. 1 will go HIGH

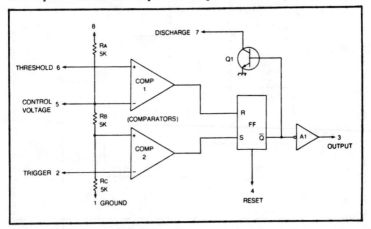

Fig. 29-12. Internal circuit of 555.

when the control voltage applied to pin No. 5 of the IC is $2/3\text{-}V_{cc}$. Similarly, the same voltage divider is used to bias comparator No. 2. The voltage applied to the noninverting input of the second comparator is given by:

$$E_2 = \frac{R_c}{R_a + R_b + R_c} \times (V_{cc})$$

This means that the output of comparator No. 2 will go HIGH when the voltage applied to pin No. 2 (trigger input) of the IC drops to $1/3\text{-}V_{cc}$. This will cause the flip-flip to go into the SET condition and cause the not-Q flip-flop output to drop LOW.

A drop to the LOW state by the not-Q output (at time T_1) causes two things to occur simultaneously. One is to force the output of the buffer amplifier (U1) HIGH, and the other is to turn off transistor Q1. This allows capacitor C1 to begin charging through resistor R1. The voltage across C1 is applied to the noninverting input of comparator No. 1 through the *threshold* terminal (pin No. 6). When this voltage reaches $2/3\text{-}V_{cc}$, comparator No. 1 will toggle to its HIGH state and will reset the RS flip-flop. This occurs at time T_2 and forces the not-Q output again to its HIGH state.

The output of amplifier U1 again goes LOW, and transistor Q1 is turned back on. Whenever Q1 is on, capacitor C1 is discharged. At this point the cycle is complete, and the 555 again in its dormant state.

The output terminal will remain LOW until another trigger pulse is received. The approximate length of time that the output terminal remains HIGH is given by the equation, $T = 1.1\ R_1 C_1$, where the resistance is in ohms, and the capacitance is in farads.

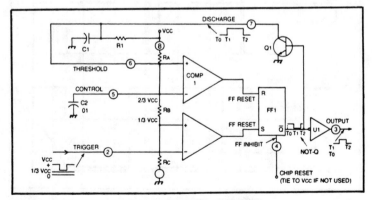

Fig. 29-13. 555 internal and external circuit for one-shot.

Figure 29-13 shows the circuit of the monostable multivibrator superimposed on the block diagram of the 555 device. This arrangement makes it easier to understand the preceding operation of the circuit.

The output waveforms and capacitor waveforms are shown in Fig. 29-14, while the graph for the timing function is shown in Fig. 29-15. This graph will help select resistor and capacitor values for time periods of 0.01 and 10.0 seconds.

If the reset terminal is not used, it should be tied to V_{cc} to prevent noise pulses from jamming the flip-flop. If, however, negative-going pulses are applied simultaneously to the trigger input (pin No. 2) and the reset terminal (pin No. 4), the output pulse will terminate. When this occurs the output terminal drops immediately back to the LOW state, even though the time period has not yet expired.

The trigger input of the 555 IC timer is normally held in a HIGH condition when the device is supposed to be dormant. To trigger the chip into producing an output pulse, we must drop the trigger input (pin No. 2) to a level of $1/3$-V_{cc} or less. Figure 29-16 shows how to make a manual trigger circuit for the 555. Resistor R4 is a pull-up resistor used to keep pin No. 2 HIGH. Capacitor C3 is charged through resistors R3 and R4. When S1 is pressed, the junction of R3 and C3 is shorted to ground, rapidly discharging C3. The sudden decay of the charge on C3 generates a negative-going pulse at pin No. 2 that triggers comparator No. 2, initiating the output pulse sequence.

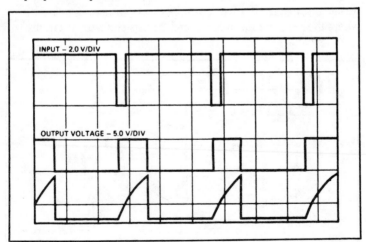

Fig. 29-14. Timing diagram of Fig. 29-13 circuit.

298

The 555 was one of the first IC timers of the comparator-RC type of design. Since the 555 was introduced, other types have been designed. One of these is a device made by *Exar* and *Intersil* called the XR-2240.

RETRIGGERABLE 555 ONE-SHOT

A retriggerable monostable multivibrator is one that will allow the output to respond to a subsequent input trigger pulse. Normally, a monostable multivibrator will not accept retriggering until the output pulse has timed, and, sometimes, a short refractory period also expires. In a retriggerable model, however, the output will be extended for one RC time period every time that a trigger pulse is received. Figure 29-17 shows a partial circuit of the 555 one-shot with two additional transistors used to allow retriggering. When a trigger pulse is received (note that the trigger pulse is positive-going), both Q1 and Q2 are turned on hard. Transistor Q1 is used to short out the timing capacitor at the time when we want to retrigger the circuit. This sets the condition back to zero, allowing it to begin again at the T_0 point. The 555 trigger input (pin No. 2) is normally held HIGH when the device is dormant by pull-up resistor R1. When the trigger pulse turns on transistor Q2, however, pin No. 2 drops LOW, allowing the device to be triggered.

EXAR XR-2240 TIMER

The XR-2240 is an RC-comparator timer IC that uses a clock section much like the 555, but it also includes an eight-bit binary counter. It is in the binary counter that we find the greatest use of

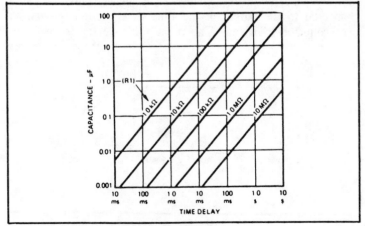

Fig. 29-15. Frequency chart of Fig. 29-13 circuit.

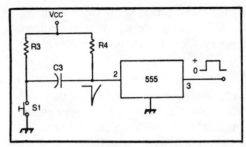

Fig. 29-16. Pushbutton triggering of the 555.

this IC. The internal circuitry, block diagram of the XR-2240 is shown in Fig. 29-18A. The timer will, similar to the 555, operate over a supply voltage range of +4.5 to +18 VDC. The time base section is a clock circuit that uses the same principles as the 555 device. This fact may be seen in Fig. 29-18B, which shows the internal circuitry of the XR-2240. One main difference between the 555 circuit and the time base section of the XR-2240 is the relative reference levels created by the internal voltage divider resistors used to bias the comparators (R1, R2, and R3). In the 555 device, all three resistors were of equal value, so we had comparator input levels of 0.667 and 0.333 times the V_{cc} potential. In the Exar chip, on the other hand, the reference levels are $0.27V_{cc}$ and $0.73V_{cc}$, respectively. One result of this is simplification of the equation that gives the period of the output waveform. In the case of the 555, the period was 1.1RC. In the XR-2240, it s simply RC—the time constant.

The binary counter portion of the time IC consists of a chain of J-K flip-flops connected in the standard manner, where each stage functions as a divide-by-2 counter. The binary counter chain is connected to the output of the time base section through an open-collector transistor. The transistor collector is also con-

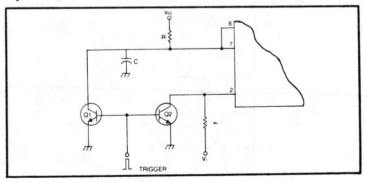

Fig. 29-17. McCullough retriggerable monostaole multivibrator (555).

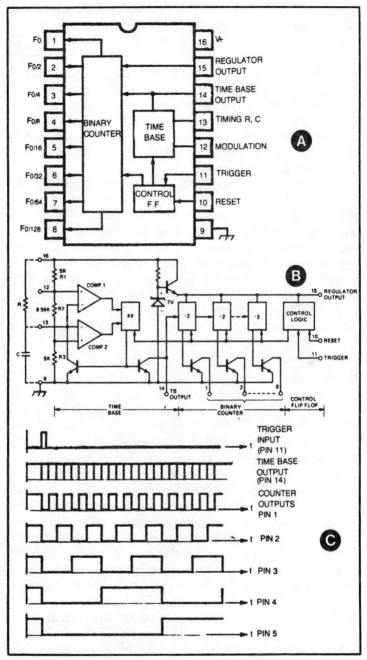

Fig. 29-18. Exar XR-2240 timer IC.

301

nected to IC pin No. 14 (called the *time base output*) so that a 20K ohm pull-up resistor can be connected between the collector and the output of the internal regulated power supply (pin No. 15).

Digital outputs from this counter are, in the usual fashion, given as voltage levels at a set of IC pins. Each output bit is delivered to a specific terminal of the IC package where it is connected to a pull-up resistor similar to that used for the time base output terminal. The output terminals will be LOW when active. This arrangement may seem opposite to the normal way of doing things, but there is a method to this madness that results in a stable, long-duration timer.

Figure 29-19 shows the basic operating circuit for the XR-2240 timer. This chip proves interesting because the sole difference between astable and monostable operation is the 51K ohm feedback resistor (R2) linking the reset terminal (pin No. 10) and the wired-OR ouput pins. The timer is set into operation by application of a positive-going trigger pulse to pin No. 11. This pulse is routed to the control logic and has several jobs to perform simultaneously: resetting the binary counter flip-flops, driving all outputs LOW and enabling the time base circuit. As was true in the 555 IC, this timer works by charging capacitor C1 through resistor R1 from the positive voltage source V_{cc}. The period of the output pulse is given by:

$$T = RC$$

where T is the time in seconds, R is the resistance in ohms, and C is the capacitance in farads.

The pulses generated in the time base section are counted by the binary counter section, and the output stages change states to reflect the binary count. This process will continue until a positive-going pulse is applied to the reset terminal.

Figure 29-18C shows the relationship between the trigger pulse, time base pulses and various output pulse states. The reason for the open-collector output circuit is to allow the user to wire a permanent-OR output so that the actual duration can be programmed. Each binary output is wired in the usual power-of-two sequence: 1,2,4,8,16,32,64 and 128. If these are wired together, the output will remain LOW as long as any *one* output is LOW. This allows us to program the output duration from $1T$ to $255T$, where T is defined as RC. We connect together the outputs required to sum the time base pulses to the desired period. For example, suppose that we wanted to design a 57-second timer.

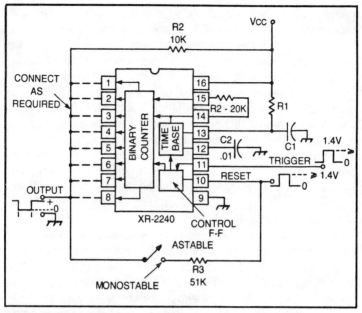

Fig. 29-19. XR-2240 circuit.

With ordinary RC one-shots, such a long time will be difficult to achieve because of errors and drift of the high-value capacitors and resistors needed. A 57-second or longer timer could be made with an XR-2240 IC, however. If the time base clock is set to 1 second (RC = 1), we could wire together outputs to sum to 57, such as 32 + 16 + 8 + 1 = 57. The time base components could be 1 megohm and 1 microfarad to make a 1-second clock.

We could change the time base, or the wired-OR terminals, to change output periods. Of course, if the time base frequency were doubled, the counter would reach the desired state in one-half the time. This feature allows programming of the XR-2240 to time durations that prove difficult to achieve with other circuitry.

Each output of the XR-2240 must be wired to the V_{cc} power supply through a pull-up resistor, unless, of course, the wired-OR output configuration is used. In that case, a single 10K ohm pull-up resistor is used. Current through the output terminals must be limited to 5 milliamperes, each. This limit will serve as a guide to the selection of suitable pull-up resistor values for any given power-supply potential.

The amplitudes of reset and trigger pulses must be at least two pn junction voltage drops: i.e, 2×0.7 volts = 1.4 volts. In most practical applications it might be wise to use a pulse of at least 4 volts amplitude, or standard TTL levels, in order to guard against the possibility that any particular chip may be a little difficult to trigger near minimum values, or that some external factor might conspire to drop pulse amplitude at some critical moment.

Synchronization to an external time base, or modulation of the pulse width, is possible by manipulating th chip through pin No. 12. In normal operation, this pin, which is the noninverting input of comparator No. 1, is bypassed to ground through a 0.01-μF capacitor. This is done to reduce the effect of noise on the operation of the timer. A voltage applied to pin No. 12 will vary the pulse width of the signal generated by the time base. This voltage should be +2 volts and +5 volts for a time base change multiplier of 0.4 to 2.25, respectively.

If you want to synchronize the internal time base to an external reference clock, connect an RC network consisting of a 0.1-μF capacitor and a 5.1K ohm resistor to IC pin No. 12. This forms an input network for the sync pulses, and these should have an amplitude of at least 3 volts with periods between 0.3T and 0.8T (see Fig. 29.20). Another way to link the count rate to an external reference is to use an external time base directly. This signal may be applied to the *time base output* terminal, pin No. 14.

Each XR-2240 has its own internal voltage regulator circuit to hold the DC potentials applied to the binary counters to a level compatible with TTL logic. This consists of a series-pass transistor which has its base held to a constant voltage by a zener diode. If operation below +4.5 VDC is anticipated, it becomes necessary to strap the regulator output terminal (pin No. 15) to V_{cc} (pin No. 16). The regulator terminal can be used to source up to 10 milliamperes to an external circuit or to another XR-2240.

LONG-DURATION TIMERS

A long-duration timer can be defined as one in which the timer period is more than about five minutes. There are several different circuit methods that can be used to make such a timer. For example, you could connect a unijunction transistor (UJT) into a relaxation oscillator circuit, or use a 555. In almost all cases, though, there seems to be an almost inevitable error created by temperature coefficient and inherent tolerance limits of the high-value capacitors and resistors needed for that service. Also, a certain amount of voltage drop is across the capacitor due to its own internal leakage resistances and the impedance of the circuit in which it is connected. The use of the XR-2240 timer will all but eliminate such problems because we are able to use the easier-to-stabilize higher clock frequencies allowed by the binary counter. A higher clock frequency is usually easier to tame because of the lower value components that are allowed. This makes the clock frequency both initially more accurate and more stable in the long run.

An example of a long-duration timer based on the XR-2240 is shown in Fig. 29-21; it uses two XR-2240 devices in cascade to increase the time duration. In this circuit, time base output of IC2 (pin No. 14) is used as an input for an external time base. Timer IC1, also an XR-2240, is used as a time base for the second timer. The next most significant bit of IC1 is triggered until time $T_0 = 128R_1C_1$ and will then go HIGH and trigger IC2.

The binary counters in IC2 will increment once every $128T$. Timer IC1 is essentially operating in the astable mode because its reset pin is tied to the reset pin of IC2, and that point does not go HIGH until the programmed count for IC2 forces its output to go HIGH.

The total time duration for this circuit under the conditions shown—IC2 input from pin No. 8 on IC1, and with all IC2 outputs connected into the wired-OR configuration—is 256 x 256, or $65,536T$. You can, however, custom program the timer to your own application by manipulating three factors: time base period

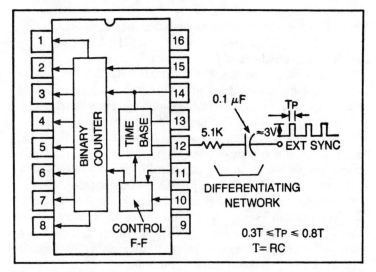

Fig. 29-20. External sync.

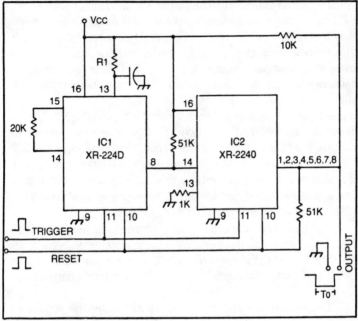

Fig. 29-21. Long-duration timer.

(R1C1), the output pin on IC1 used to trigger IC2, and the strapping configuration on IC2. In other words:

$$T_o = R_1C_1 \times T_0'' \times T_o''$$

where T_o is the total period in seconds that the output is LOW, T'' is the total time period the selected output o IC1 is LOW, and T_o'' is the total time weighting of the IC2 output.

As an example, assume that the product R1C1 is one second (R1 = 1 megohm, C1 = 1 uF). The output remains LOW for a period of T_o = 65,536T, or (65,536)(R1C1) = 56,536 seconds (about 18 hours). Of course, it is almost impossible to accurately generate a time delay this long using any of the other techniques. It is, for example, relatively common to find high-value, electrolytic capacitors—necessary in long-duration RC timers—rated with a − 20 percent to +100 percent tolerance in capacitance. This is incompatible with the goal of making a precision time base of long duration. Another complication is that electrolytic capacitors— even the tantalum types—tend to change value with time. This situation would be tolerable in most filtering or decoupling applications, but ruinous in a timer application.

Chapter 30
RC Square Wave
and Triangle Oscillators

Oscillator circuits that were introduced into an early chapter produce sine wave output signals. In some cases, though, we will want to generate some other waveform, including sawtooth, square wave and triangle. Oscillator circuits that produce these waveforms are covered in this chapter.

Such waveforms are in a general class called *nonlinear waveforms*. The sine wave can be considered linear because it has a Fourier series that includes only the fundamental. As the waveform departs from the pure sinusoidal shape, however, its frequency component content begins to increase. Any such waveform can be represented as a series of harmonically related sine or cosine waves of varying amplitudes or phases. These are the harmonics that are found in nonlinear waveforms. Clearly, different techniques will be used to generate these waveforms, compared with sine wave oscillator circuits.

SAWTOOTH GENERATORS

A sawtooth generator is a circuit that will produce a ramp output waveform (Fig. 30-1). The voltage in a positive-going sawtooth will start from some initial voltage (usually zero), rise to a maximum voltage, and then drop back to the initial level. The name of this waveform is derived from the fact that a train of them displayed on an oscilloscope resembles the teeth of a saw. There are two ways to generate a sawtooth with analog components: *unijunction transistor relaxation oscillator* or the *Miller integrator*.

The relaxation oscillator is shown in Fig. 30-1B. This circuit uses the unijunction transistor (UJT), although some earlier versions used neon glow lamps such as the NE-2 or NE-51. The UJT consists of a channel and an embedded emitter junction. The two ends of the channel are labeled base-1 (B1) and base-2 (B2). The emitter forms a pn junction with the base channel. Recall from transistor theory that the pn junction will not conduct current (except for a tiny leakage current) until there is an appropriate bias voltage across the junction. For germanium junctions the magic number is 0.2 to 0.3 volts, while for silicon devices it is 0.6 to 0.7 volts. When the base-emitter potential exceeds this level, the junction will break over and present a low resistance.

The junction potential in Fig. 30-1B is controlled by the potential across capacitor C, which is in turn controlled by the RC time constant of the capacitor and its series resistor R. The capacitor will continue to charge as long as the voltage across the capacitor is less than the breakover voltage of the UJT. When the capacitor voltage reaches the breakover voltage, however, the pn junction will suddenly conduct current and will discharge the capacitor rapidly. The output of the relaxation oscillator is taken across the capacitor and will be as shown in Fig. 30-1B. This is unfortunately not an ideal sawtooth because the leading edge of the waveform is the exponential curve one expects from a capacitor charge waveform.

A solution to the problem is shown in Fig. 30-1C. In this UJT relaxation oscillator, the resistor used to charge the capacitor has been replaced with a constant current source (CCS). This might be a JFET diode connected as a CCS, or a pair of bipolar transistors configured to be a CCS. In either event, the circuit will produce a

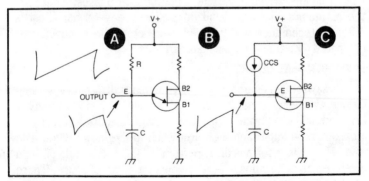

Fig. 30-1. Sawtooth waveform (A) semisawtooth UJT circuit (B) and improved version (C).

constant current level, regardless of changes in the load, or the stage of charge on capacitor C. The result of using a CCS to charge the capacitor is to linearize the leading edge ramp of the sawtooth waveform. The amended waveform is also shown in Fig. 30-1C.

A Miller integrator circuit is shown in Fig. 30-2A. This circuit is the standard operational amplifier integrator with a means for automatically terminating the integration process.

In the Miller integrator circuit used for sawtooth generation, we must provide a means for charging the capacitor in the operational amplifier feedback loop with a constant current. Since the noninverting input is grounded, the inverting input is at virtual ground. Input resistor R is fixed, so we will generate a constant current, E_{ref}/R, if we connect a constant reference voltage to the input side of the resistor. Current I1, then, is constant.

The waveform of the Miller integrator sawtooth generator is shown in Fig. 30-2B. The trip level of the comparator (E1) sets the maximum amplitude of the output sawtooth. Recall that a comparator is a circuit that will produce an output that indicates whether two voltages are equal, or in which polarity they are unequal. As long as E_o is less than bias level E1, the output of the comparator will be zero. When the output potential rises to become equal to E1, however, the output of the comparator jumps HIGH and will turn on CMOS electronic switch S1. This switch has the job of dumping the charge in capacitor C1, thereby returning the output voltage to zero. The leading edge of the sawtooth will be linear because the charging current applied to the operational amplifier input is constant. When the output voltage drops below E1, switch S1 opens again, permitting the beginning of a new cycle.

We will mention one last method of generating a sawtooth waveform, although we will not go into it very deeply. A digital-to-analog converter (DAC) and a binary counter can generate a staircase waveform. If the step increments are sufficiently small, the waveform will behave much like a sawtooth ramp. The DAC produces an output voltage (or, in some models, current) that is proportional to a DC reference and a binary word applied to its digital inputs. If the digital inputs of the DAC are connected to the binary outputs of a digital counter, the DAC output will ramp upward as the count progresses, resetting to zero when the counter overflows.

SQUARE WAVE OSCILLATORS

A square wave oscillator can be made quite simply. In many ways, it will perform better than some of the sine wave oscillator circuits. One major area of superior performance is the matter of amplitude stability. It is sometimes very difficult to build a sine wave oscillator that will maintain a constant amplitude output signal. The problems, however, are considerably lessened in most square wave designs.

One quick method for building a square wave oscillator is to use the 555 integrated circuit timer (Fig. 30-3). This device was also used in the creation of certain monostable multivibrator circuits in Chapter 29, but here we are dealing with the *astable* multivibrator. Just what is an astable? This word means that it has no stable states. It will flip back and forth between the two possible binary states but will not rest at either of them. The output, therefore, will be a square wave.

The block diagram of the internal circuitry of the 555 is shown in Fig. 30-3. The principal stages are two comparators, an RS

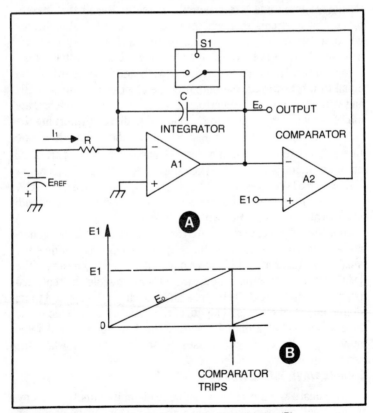

Fig. 30-2. Miller integrator sawtooth generator (A) and timing (B).

310

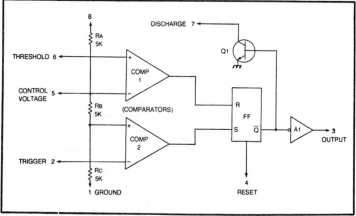

Fig. 30-3. Internal circuit for the 555.

flip-flop, an output inverting amplifier (A1) and a discharge transistor (Q1). The 555 is of bipolar design, but it is not TTL or any of the other standard logic types. It will operate with V+ potentials between 4.5 and 15 volts (some versions to 18 volts), and it is timed by an RC network.

Figure 30-4A shows the 555 connected into its astable multivibrator circuit, with the circuit repeated in Fig. 30-4 (showing only the external connections). We will discuss the circuit in terms of the block diagram, so use Fig. 30-4A.

The bias levels on the comparators are set by a voltage divider network, R_a, R_b and R_c. All three of these resistors have the same value, nominally 5000 ohms. This means that comparator No. 1 is biased to $(2/3)V_{cc}$ and comparator No. 2 is biased to $(1/3)V_{cc}$. Comparator No. 1 is used to reset the flip-flop, while comparator No. 2 is used to set the flip-flop. Note that the inverting input to comparator No. 2 is specified as the trigger input. The 555 output pulse is initiated by having the trigger input drop to less than $(1/3)V_{cc}$. Since both the threshold and the trigger inputs are connected across the timing capacitor, the circuit is a self-retriggering multivibrator; hence, astable operation is obtained.

The circuit of Fig. 30-4A is specifically a 555 version of the astable multivibrator concept. The voltage applied to the two external comparator inputs (threshold and trigger) is determined by the time constant, C1(R1+R2). Under initial conditions, the not-Q output of the RS flip-flop is HIGH. This will turn on transistor Q1, keeping the junction of resistors R1 and R2 at ground potential. Capacitor C1 has been charged, but when Q1 is turned

311

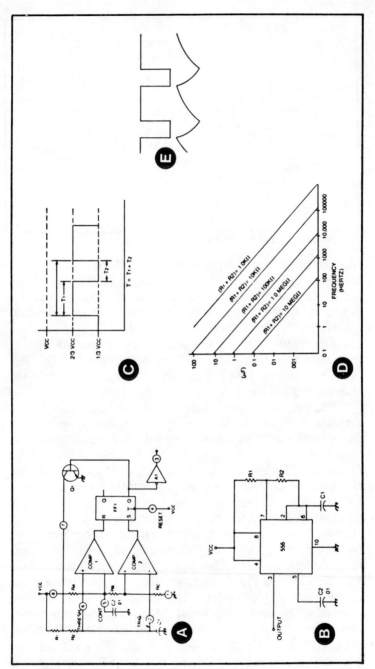

Fig. 30-4. 555 as an astable mutivibrator.

on, it will discharge C1 through resistor R2. When the voltage across capacitor C1 drops to a level of $(2/3)V_{cc}$, the output of comparator No. 1 goes HIGH and resets the flip-flop. This action again turns on Q1 and allows C1 to discharge to $(1/3)V_{cc}$. Capacitor C1, then, alternatively charges to $(2/3)V_{cc}$ and discharges to $(1/3)V_{cc}$. Figure 30-4C shows the relationship in the output waveform between HIGH and LOW states. The HIGH time, t_1, is given by:

$$t_1 = 0.693(R1 + R2)C1 \qquad \text{Equation 30-1}$$

and the low time by:

$$t_2 = 0.693(R2)C1 \qquad \text{Equation 30-2}$$

The total period of the waveform, T, is the sum of t_1 and t_2, and is given by:

$$T = t_1 + t_2 \qquad \text{Equation 30-3}$$

$$T_{sec} = 0.693(R1 + 2R2)C1 \qquad \text{Equation 30-4}$$

The frequency of any oscillation is the reciprocal of its period, so we can find the operating frequency of the circuit from Equation 30-4:

$$f_{Hz} = \frac{1}{0.693(R1 + 2R2)C1} \qquad \text{Equation 30-5}$$

$$f_{Hz} = \frac{1.44}{(R1 + 2R2)C1} \qquad \text{Equation 30-6}$$

Equation 30-6 is the principal equation used to design 555 astable multivibrators. This equation is solved graphically for frequencies between 0.1 and 100,000 Hz in Fig. 30-4D. The solution set chosen permits selections of easily obtained component values.

The relationship between the C1 potential and the output stage is shown in Fig. 30-4E. The duty cycle, also called duty factor, is the percentage of the total on period that the output is HIGH. This is given by the expression:

$$\text{D.F.} = \frac{t_1}{t_1 + t_2} = \frac{R2}{R1 + R2} \qquad \text{Equation 30-7}$$

We can, therefore, vary the duty cycle of the 555 astable multivibrator by adjusting the ratio of the two resistors used in the timing network. The actual frequency is dependent upon the total resistance (see Equation 30-6), but the on-time of the output square waves is dependent upon the ratio of R1 and R2 (as defined in Equation 30-7).

Numerous integrated circuits will produce a square wave output, but there is space for only a few of them here. Keep up with the offerings of the various semiconductor manufacturers by consulting their respective catalogs and applications notes.

In most of the astable circuits of this chapter, we will use the operational amplifier. This device works nicely, is reasonably universal and amply demonstrates some of the principles of operation with which we are concerned.

OPERATIONAL AMPLIFIER ASTABLE

Figure 30-5A shows the circuit for the basic operational amplifier astable multivibrator. The timing for this circuit is controlled by resistor R3 and capacitor C1. The operational amplifier works as a comparator in this circuit. One input (noninverting) is connected to a voltage divider that produces a potential that is a fraction $(R2/(R1+R2))$ of output potential E_o. The other input of the comparator (inverting) is connected across the voltage at point A, the capacitor charge voltage.

The timing diagram is shown in Fig. 30-5B. The output will be at V+ initially, and that capacitor C1 is charged to V−. The output remains at V+ as long as capacitor voltage E_{c1} is less than the voltage at point B (V_B). When the capacitor voltage reaches V_B, however, the comparator output switches, making the output potential V−. The charging current for the capacitor is now reversed due to the change in E_o polarity, so it begins to discharge the capacitor. The discharge cycle will continue until the capacitor voltage reaches $-V_B$. At that time, the cycle will again reverse itself, and one complete square wave will have been completed. The period of the output waveform is T, which is found from:

$$T = 2R_3C_1 \times \ln \left[\frac{2R_1}{R_2} + 1 \right] \qquad \textbf{Equation 30-8}$$

where T is the time in seconds, all resistances are in ohms, all capacitances are in farads, and ln denotes the base of the natural logarithms.

The circuit in Fig. 30-5A is capable of operating with ease over the range of 1 Hz to 10 kHz and can be extended to higher and lower limits by clever circuit design and operational amplifier selection. The output voltage limits are set by a pair of back-to-back zener diodes across the output terminal. The output waveform is symmetrical if the zener potentials of D1 and D2 are matched exactly. An asymmetric output waveform is possible by selecting different zener potentials for D1 and D2. It is generally true that the use of the zener diodes will make the corners of the output waveform sharper than could be obtained with just the operational amplifier alone. If the two diodes are identical, we can expect an output voltage of $V_z + 0.7$ volts. The 0.7V factor is due to the effect of the forward biased diode of the pair on each half-cycle.

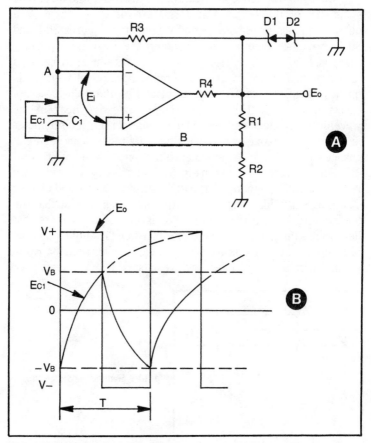

Fig. 30-5. Operational amplifier square wave generator (A) and waveform (B).

315

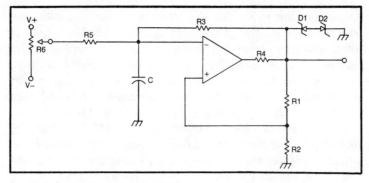

Fig. 30-6. Symmetry change circuit.

Changing the diode configuration will change the amplitude symmetry of the output waveform. We must, however, use different tactics to provide period asymmetry. Figures 30-6 and 30-7 show two methods for providing the required symmetry changes. In Fig. 30-6, a potentiometer network similar to that used in operational amplifier circuits to null output offset voltages is employed. In this case, a bias is placed on the voltage across capacitor C1. Whether the positive or negative peak of the waveform is changed will depend upon whether the potentiometer is set to the negative or positive side of its range. The degree of change will depend upon the actual voltage on the wiper of R6.

Another tactic is shown in Fig. 30-7. In this case, however, we require no potentiometer, but use a pair of resistors and diodes. The diodes act like polarity-sensitive switches. Recall that the feedback resistance (Equation 30-8) is one of the factors that determines the period of the waveform. When the output potential is positive, diode D4 is forward biased, placing resistor R6 in parallel with the regular feedback resistance R3. The resistance

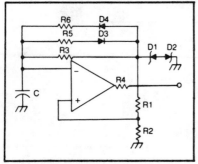

Fig. 30-7. Symmetry change circuit.

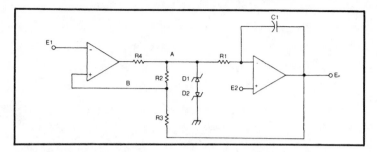

Fig. 30-8. Triangle generator.

that must be used in calculating the time that the output period is high is the combined parallel resistance. Similarly, on the negative excursion of the output waveform, diode D3 is forward biased, placing resistor R5 in parallel with resistor R3. If we keep the two diode-controlled resistors unequal, the symmetry of the positive and negative halves of the waveform is destroyed.

TRIANGLE GENERATOR

The operational amplifier comparator, coupled with a Miller integrator, can form a triangle waveform generator. The output from an integrator, which the input signal is a square wave, will be a triangle. The voltage at point "A" in Fig. 30-8 is a square wave whose frequency is controlled by R1 through R3 and capacitor C1. The frequency, then, is expressed by:

$$f = \frac{R_2}{4 R_3 R_1 C_1} \qquad \textbf{Equation 30-9}$$

The inverting input of the comparator stage is biased by a reference voltage, while the other input receives a bias from both the comparator and the integrator output terminals. We can calculate the two peak voltages from the expressions:

$$E_{o(max)} = \frac{E1(R2 + R3)}{R2} + \frac{E_A R3}{R2} \qquad \textbf{Equation 30-10}$$

$$E_{o(min)} = \frac{E1(R2 + R3)}{R2} + \frac{E_A R3}{R2} \qquad \textbf{Equation 30-11}$$

From these expressions, the peak-to-peak voltage of the output triangle waveform can be derived:

$$E_{p\text{-}p} = E_{o(max)} - E_{o(min)} = 2E_A R_3 / R_2 \qquad \textbf{Equation 30-12}$$

317

Chapter 31
Crystal Oscillators

Oscillators must have some means of discriminating against unwanted frequencies and still produce the desired frequency. Some circuits are phase shift oscillators, while others are feedback oscillators. In the feedback type, we must provide a tuned circuit that selects the desired frequency. In some cases, the tuned circuit will be a series-tuned or parallel-tuned circuit using inductor and capacitor elements. These circuits are discussed in chapter 28 (LC oscillator circuits). But LC elements have certain problems. They are subject to changes in resonant frequency due to vibration, changes in temperature and the hot breath of pure fickle fate. Some old, low-cost transmitters, usually amateur radio models, had LC variable frequency oscillators that would shift frequency when someone walked into the operating room. When a high degree of frequency control and freedom from some of the faults of the LC oscillator are desired, we might opt for the *piezoelectric crystal oscillator*. Before we study the piezoelectric crystal oscillator, though, let us first examine the piezoelectric crystal resonator element.

PIEZOELECTRIC CRYSTAL ELEMENTS

Some crystalline elements, notably quartz, possess a property called *piezoelectricity*. This phenomenon is not found in all materials, only certain crystals and ceramic materials. *Piezoelectricity* refers to the generation of electrical potentials when the material is *deformed*. When a piezoelectric element is mechanically deformed, therefore, a voltage appears across the faces of the crystal slab. There is also an inverse phenomenon: When a voltage is applied to the faces of the crystal slab, the crystal will mechanically deform. We can use these phenomena to cause the crystal to oscillate. If a pulse is applied to the faces of the crystal slab, then, the slab will vibrate back and forth in an oscillatory manner. Losses in the slab will cause the amplitude of these mechanical oscillations to die out in a manner that suggests exponential decay. While the pulsed crystal slab is oscillating, it will generate an AC voltage across the faces of the slab that has the same frequency as the mechanical vibrations.

Figure 31-1A shows the circuit symbol for a piezoelectric crystal resonator, while Fig. 31-1B shows the mechanical form the actual crystal will take. The cut slab of crystal material is sandwiched between two contact electrodes. In older crystals, these electrodes are attached with spring tension, while in modern crystals they are silver deposited directly onto the surface of the slab. The crystal element is mounted to the pins protruding through the header (a support structure).

Figure 31-1C shows the equivalent circuit for a piezoelectric crystal element. A series resistance, series inductance and series capacitance are in the circuit. There is also a parallel capacitance in the circuit. The frequency response of a crystal element is shown in Fig. 31-2. This graph plots the reactance-versus-frequency. Note that there are actually two resonant modes for the crystal: *series* and *parallel*. You would have guessed that from the fact that there are two capacitors that interact with the inductor of Fig. 31-1C. One of the capacitors (C_s) forms a *series* resonant circuit with inductor L_s, while the other capacitor (C_p) forms a *parallel* resonant circuit with inductor L_s.

The series resonant frequency is the frequency at which the inductive reactance (X_{L_s}) of the series inductor exactly cancels the capacitive reactance. At this point, the total reactance of the crystal is zero; only the series resistance, R_s, determines the crystal impedance. The impedance of the crystal is, then, at a minimum for the series resonant frequency.

The parallel resonant frequency of the crystal is 1 kHz to 15 kHz higher than the series resonant frequency. The impedance of the crystal at the parallel resonant frequency is maximum (see Fig. 31-2). Note that a parallel-mode crystal will operate in the series mode if a small series capacitor is included. The value of the capacitor must be equal to the specified load capacitance of the crystal. The opposite, however, is not usually true.

There are two oscillatory modes for crystal resonators: *fundamental* and *overtone*. The fundamental frequency is the *natural* resonant frequency, or the frequency at which the slab will oscillate if stimulated. Fundamental-mode oscillation depends upon the mechanical dimensions of the crystal, the style of cut and several other factors.

In the overtone mode, the crystal oscillates at a frequency that is approximately an integer multiple of the fundamental frequency. Note, however, that the overtone is *not* a harmonic of the

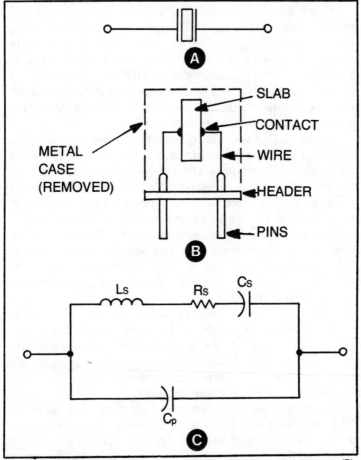

Fig. 31-1. Symbol for piezoelectric crystal (A), crystal resonator structure (B) and equivalent circuit for a crystal (C).

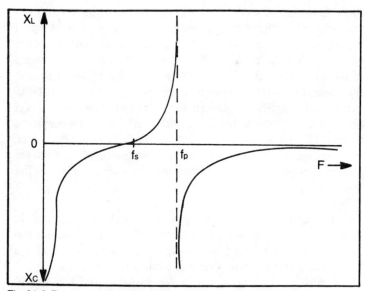

Fig. 31-2. Frequency response of equivalent crystal circuit.

fundamental frequency. If you divide the fifth overtone frequency by five, the resultant frequency is not the fundamental frequency, but a frequency close to the fundamental. The overtone frequencies are always odd (third, fifth, seventh, ninth and 11th), and the overtone crystal always operates in the series mode.

When ordering crystals, the frequency of operation and—if the parallel resonance is selected—the load capacitance must be specified. The operating frequency of the fundamental mode is easily understood. The frequency marked on the case is the fundamental frequency. Note that this frequency will be less than 20 MHz. At frequencies higher than 20 MHz, the crystal becomes too thin for safe operation and may fracture under normal operation. Fundamental crystals are usually parallel resonant types, except below 500 kHz where it is usually the practice to order series resonant crystals. The parallel resonant condition, incidentally, produces a 180-degree phase shift at the parallel resonant frequency.

The power dissipation of the crystal must be limited to prevent fracturing the slab. It is the equivalent series resistance (ESR) of the crystal (see Fig. 31-1C) that determines the power dissipation for any given level of applied signal. In a practical oscillator circuit, the dissipation is varied by controlling the feedback signal supplied to the crystal by the amplifier element.

Most fundamental-mode crystals will dissipate up to 200 microwatts, although good engineering practice is to make the actual dissipation somewhat less than this maximum. Low-frequency crystals—those below 1000 kHz—usually have a dissipation rating of 100 microwatts. It is generally regarded as good practice to limit the dissipation to 50 microwatts to improve the frequency stability of the crystal oscillator. We do not try to obtain large amounts of rf power from a crystal oscillator although some amateur radio projects of a few years ago took that approach. We limit the power of the crystal oscillator in order to improve the stability. Then we use an rf power amplifier to increase the output power of the system.

CIRCUITS

The crystal element is the equivalent of an LC tuned circuit. In fact, it is equivalent to two resonant tank circuits (parallel and series). We must connect the crystal into an oscillator circuit in a manner similar to the place occupied by the LC tank circuit.

Figure 31-3 shows a circuit for a simple *100-kHz crystal calibrator*. These circuits are often used by amateur radio operations, shortwave listeners and other operators of shortwave receivers to calibrate the dial pointer. The oscillator operates at a fundamental frequency of 100 kHz, and the harmonics are used to locate points on the shortwave dial, provided that the output of the calibrator is coupled to the antenna circuit of the receiver. The active element in this circuit is a bipolar npn transistor. A 2N2222

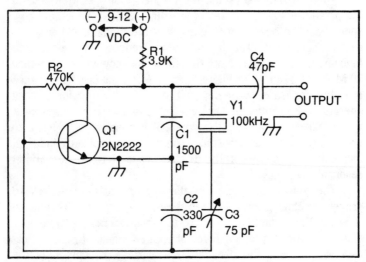

Fig. 31-3. A 100-kHz crystal calibrator.

is selected, but this is not too critical. Almost any transistor that provides good gain (50 to 100) with an f_t of 50 MHz or more will work in this circuit. The 2N2222 was selected because it is easily obtained, and replacements from the "blister pack" manufacturers are usually available in all localities.

Note that the circuit configuration for the 100-kHz crystal calibrator circuit is the familiar Colpitts oscillator. Feedback is a function of a capacitor voltage divider consisting of C1 and C2. The ratio of these capacitors is a trade-off between output amplitude and stability, and it also sets the power dissipation of the crystal element. Collector load is provided by resistor R1, and the transistor is biased in the collector-base manner by resistor R2.

The crystal element (Y1) shunts the feedback voltage divider, and is in series with a variable capacitor (C3). This capacitor is used to set the actual operating frequency of the calibrator. Almost any crystal operating frequency can be *pulled* a little bit by a series or parallel capacitor. The output signal is coupled to the load through capacitor C4. This circuit should be lightly coupled to the load (usually a receiver antenna circuit) in order to minimize changes in frequency caused by changes in the load impedance. If this is a problem, an emitter follower output buffer stage should be used between the oscillator and the load.

Calibration of the crystal oscillator can be done in either of two ways. If you have a counter, simply measure the output frequency of the oscillator and adjust C3 until the frequency is exactly 100.00 kHz. If no counter is available, but a shortwave receiver *is* available, then use the standard frequency broadcasts of the National Bureau of Standards radio stations, WWV and WWVH. WWV is located at Fort Collins, Colorado, while WWVH is in Hawaii. These stations transmit signals on precisely controlled frequencies of 2.5 MHz, 5.0 MHz, 10.0 MHz, 15.0 MHz and 20.0 MHz. Select the highest WWV or WWVH frequency that provides a strong signal at your location. Then zero beat the crystal frequency (using C3) to the WWV/WWVH signal. Note that this job should not be done until the oscillator has operated for at least 10 minutes. This will allow any thermal changes to come to rest before any attempt is made to calibrate the frequency (temperature is one of the variables that tends to affect the operating frequency of a crystal).

An example of a field-effect transistor version of the old Miller oscillator is shown in Fig. 31-4. This circuit was originally popularized in vacuum-tube form, but this version uses a junction

FET. The crystal is connected to the JFET gate circuit and is in parallel with a 10-megohm load resistor. Bias for the JFET is set by source resistor R2. The source terminal of the JFET is bypassed to ground for rf by capacitor C4. The value of this capacitor is selected to have a low (one-tenth or less) reactance compared with the resistance of R4 at the lowest frequency of operation.

The drain of the JFET Miller oscillator is tuned to the resonant frequency of the crystal by an LC tank circuit. Actually, the resonant frequency of the tank circuit only approximates the crystal frequency. There will be different properties to the tuning characteristic of the tank as the crystal frequency is approached from above or below. The correct setting for the tuning slug of L1 is

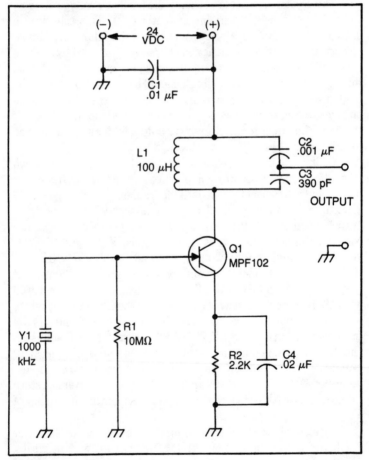

Fig. 31-4. Miller oscillator.

about 75 percent up the slope of the shallower characteristic; i.e., the side of resonance where the drain current of the FET changes least per turn of the tuning slug in L1.

Again, we must lightly couple the output of this oscillator to the load there will be frequency changes with changes of load impedance. If the load variations become a problem, use a bipolar transistor buffer amplifier (an emitter follower) between the output of the JFET Miller oscillator and the load.

The Miller oscillator in Fig. 31-4 is set up for use as a 1-MHz (1000-kHz) crystal oscillator, so it can be used as a calibrator. The frequency of this calibrator is, therefore, 10 times higher than the previous circuit. Sometimes, crystal oscillators have various frequencies in the same box, allowing some selection. If the circuit of Fig. 31-4 is followed by either a zero-crossing detector or a TTL Schmitt trigger (or any other circuit that will square up the output signal, making it compatible with TTL or CMOS devices), the circuit can be used to drive digital frequency divider ICs to produce output frequencies that are exact submultiples of the operating frequency. A single J-K flip-flop, for example, will make the operating frequency divide by two. Similarly, the 7490 TTL device can be used as a divide-by-2, divide-by-5, or divide-by-10 (the usual mode) device. A chain of 7490s in cascade will produce output frequencies of 100 kHz, 10 kHz, 1 kHz, 100 Hz, 10 Hz, 1 Hz, 0.1 Hz, etc., from the 1000-kHz oscillator.

Another example of a crystallized Colpitts oscillator is shown in Fig. 31-5. This circuit uses a slightly different, but still valid, configuration compared with Fig. 31-3. This circuit will operate with fundamental-mode crystals in the range of 1 MHz to 20 MHz—although operation, especially starting, becomes poor above 15 mHz. Feedback is controlled by capacitor voltage divider C2/C3. The rf voltage across the emitter resistor provides the basic feedback signal, which is coupled into the base of the transistor and to the crystal. Output from this stage is taken in the manner of an emitter follower. Even so, it is necessary to prevent heavy loading or wide changes in load, or there will be frequency shifting from this circuit. Again, an emitter-follower buffer amplifier will help solve some of the problems that might otherwise occur. The frequency of oscillation is set by series capacitor C1. This capacitor is adjusted to make the frequency of operation exact. Although this circuit may not be optimum, it is very popular and is well behaved for the home experimenter.

An overtone oscillator (Fig. 31-6) produces a frequency that is almost an odd-order integer multiple of the fundamental frequency.

The permissible overtones are third, fourth, fifth, seventh, ninth and 11th. The case of an overtone crystal will carry markings as to the *overtone* frequency of oscillation, *not* the fundamental frequency.

The crystal element in this circuit is connected directly between the base and ground. Capacitor C1 is used to improve the feedback due to the internal capacitances of the transistor. It might not be needed in some transistors, but don't count on it! This capacitor should be mounted as close as possible to the case of the transistor. The LC tank circuit in the collector of the transistor is tuned to the overtone frequency of the crystal. The emitter resistor capacitor must have a capacitive reactance of approximately 90 ohms at the frequency of operation. Typical values for the capacitance are given in Table 31-1.

The tap on inductor L1 is used to match the impedance of the collector of the transistor. In most cases, the optimum placement of this tap is approximately one-third from the cold end of the coil. The placement of this tap is a trade-off between stability and maximum power output. The output signal is taken from a *link coupling* coil, L2, and operates by transformer action.

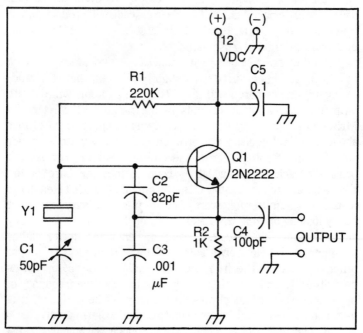

Fig. 31-5. Colpitts oscillator.

Table 31-1. Frequency-versus-Capacitance Conversion.

Frequency (MHz):	50	144	220	450
Capacitance (pF):	36	12	8.2	3.3

Overtone oscillators are typically found in VHF and UHF receivers and transmitters. At frequencies below VHF, fundamental-mode oscillators are often used. In fact, it is probably almost impossible to find a transmitter in the low end of the high-frequency range (3 MHz to 30 MHz) that uses overtone circuits. Some CB models do, but they are in the upper half of the HF spectrum. Many transmitters and receivers will place a small capacitor in series with the crystal in order to pull the frequency exactly on frequency. Note that some transmitters use fundamental-mode oscillators, and then a chain of frequency multiplier amplifiers follow to increase the frequency of the oscillator by integer amounts.

The Pierce oscillator has been around since vacuum-tube days and is characterized by having the piezoelectric crystal element placed between the output and input of an inverting amplifier. In vacuum-tube days, this meant placing the crystal between the high-voltage anode and the low-voltage control grid. In that case, a series DC-blocking capacitor was needed to keep the high voltage on the anode from fracturing the crystal. Power dissipation, always a problem in crystal oscillators, was controlled mostly by adjusting the anode potential to a low value. A trade-off was necessary, however, between the low voltage required for proper operation of the crystal and a voltage that was high enough to insure proper operation of the oscillator.

Figure 31-7 shows a solid-state version of the Pierce oscillator using pnp bipolar transistors. The oscillator is transistor Q1, and the crystal is placed between the collector and base terminals. Note that the low potential transistor circuit does not require a series capacitor to be placed in the crystal circuit. Bias for the transistor is through resistor R1 to the V− line. The base of the transistor is kept at ground potential by capacitor C1, which is also part of the feedback network. Feedback is improved by the use of the collector-emitter capacitor C2—a common feature in bipolar transistor oscillators.

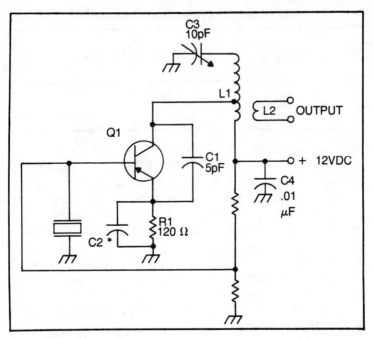

Fig. 31-6. Overtone oscillator.

Transistor Q2 is used as an output buffer. The purposes of the output buffer are: impedance transformation to low-impedance loads (typically less than 100 ohms in rf applications) and isolating the oscillator from changes in the load that could cause frequency shifts. Most oscillator circuits are not "happy" driving a variable impedance load, so some sort of buffer amplifier should be provided. The emitter-follower circuit selected for this application is ideal because it provides both the low impedance output and the isolation required. This circuit is biased to near class-A operation by resistor voltage divider R4/R5. The base of the buffer amplifier is isolated from the DC voltage on the collector of transistor Q1 by blocking capacitor C3. Note that capacitor C3 is also used in conjunction with capacitor C4 to form a capacitor voltage divider that produces an impedance match between the collector of the oscillator transistor and the base of the buffer amplifier transistor.

Another example of a Colpitts oscillator is shown in Fig. 31-8. This circuit is similar to the circuit considered earlier and will operate over the same wide frequency range. Bias for the pnp bipolar transistor is provided by resistor voltage divider network R1/R2. The collector of the oscillator transistor is kept at AC ground potential by capacitor C5. This capacitor should be placed closed to the transistor.

Feedback is provided by capacitor voltage divider C2/C3. The tapped capacitor, incidentally, is the sign of the Colpitts oscillator—something that you will find worth remembering if you face FCC or school examinations in the future. The crystal and frequency-setting trimmer capacitor (C1) are placed in shunt with the feedback network. Sometimes, the positions of the crystal and the capacitor are reversed, but the configuration shown is usually the most practical because of mounting problems with the capacitor; most trimmers, unless they are designed to be mounted on printed circuit boards, are automatically grounded when they are bolted to a chassis.

Like the earlier circuit, this circuit takes the signal from the emitter terminal through a coupling capacitor. This circuit differs from the earlier circuit, however, in that an rf choke (L1) is placed in series with the emitter resistor. This inductor will increase the rf voltage available at the output.

An example of a *Butler aperiodic oscillator* is shown in Fig. 31-9. This circuit is popular in crystal calibrators and other applications where a high harmonic content in the output waveform is desired. In this circuit, we will find that the harmonic content is very high. Indeed, the second and third harmonic amplitudes are almost as large as the fundamental amplitude. The harmonic

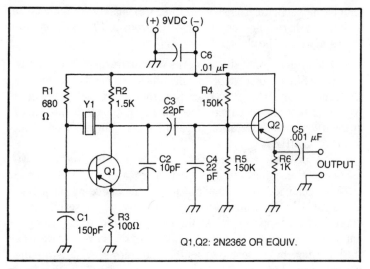

Fig. 31-7. Pierce oscillator.

329

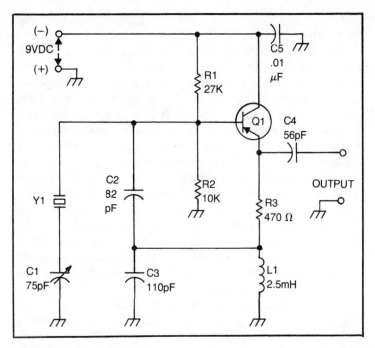

Fig. 31-8. Colpitts oscillator.

content of the output waveform can be made even richer by decreasing the value of resistor R6 (in the emitter of Q2, which is one of the oscillator transistors) to approximately 1000 ohms. In that case, harmonics up to 30 MHz from a 100- kHz crystal frequency would be obtained.

The Butler circuit shown works well in the range of 50 kHz to 500 kHz. Slight component modifications are needed for higher frequency operation. For operation over 3000 kHz, select a transistor that provides moderate gain (H_{fe} in the 60 to 150 range) at the frequency of operation and an f_t (gain-bandwidth product) of at least 100 MHz.

The crystal is placed between the emitters of the two oscillator transistors. In this particular case, a parallel-mode crystal—confused into "thinking" it is series mode by capacitors C2 and C3— is used. If want to actually use a series-mode crystal, these capacitors should be shorted out. By using a parallel-mode crystal in a series-mode circuit, however, we gain the advantage of being able to set the frequency precisely with the variable capacitor.

Diodes D1 and D2 serve to limit the output amplitude of the oscillator and to insure proper starting of the oscillations. Amplitude limiting serves to limit the power dissipation of the crystal; hence, it will improve the chances of the crystal surviving for a long time. It also tends to help the stability of the frequency of operation.

The output stage is an emitter follower which serves as a buffer amplifier. Although the bias constants and transistor type are different from circuits shown previously, the function is identical. You could, incidentally, adapt this circuit to other oscillators and thereby improve performance of some marginal circuits.

You may also want to use some of these circuits with transistors of the opposite polarity. Often npn transistors can be substituted for pnp, and vice versa, if the polarity of the power supply is reversed. This doesn't always work properly, but it works often enough to be worth a try. You might also try to substitute n-channel and p-channel field-effect transistors using the same power supply reversal trick.

The oscillator circuit of Fig. 31-10 shows a *parallel-mode aperiodic crystal oscillator* circuit that is related to the Butler circuit. In this case, the crystal is placed between the collector of the output stage and the base of the input stage.

One of the criterion for oscillation is that the signal be fed back in phase with the input signal. The amplifiers produce a 180-degree phase shift because Q1 is an emitter follower with 0-degree phase

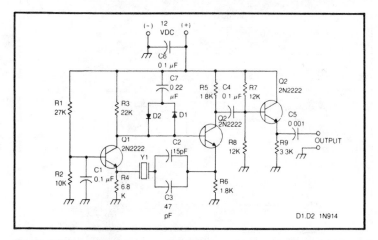

Fig. 31-9. Butler oscillator.

shift and Q2 is a common emitter stage with 180-degree phase shift. The amplifier section, therefore. is an inverter. But where does the additional 180 degrees come from? A total of 360 degrees of phase shift is needed at the frequency of operation to make this circuit oscillate. The additional 180 degrees of phase shift come from the crystal. A parallel-mode crystal produces a phase shift of 180 degrees between its electrodes *only* at the frequency of operation. At all other frequencies, the phase shift is different from 180 degrees, so there will be no spurious modes of oscillation in this circuit. The frequency of oscillation can be set to a precise value with trimmer capacitor C1. The range of operation for this circuit is 500 kHz to 10 MHz. We can extend the range downward (100 kHz) by increasing the value of C1 to 75 pF and increasing the value of C2 to 22 pF.

Digital electronics has become a big business in the past two decades. Most digital circuits—and all computers—operate from a master clock. This clock is nothing more than an oscillator with a reasonable short-term stability. Such digital circuits operate *synchronously*, meaning that changes take place only when the clock pulse is active (usually defined to mean HIGH).

There are several approaches to making a digitial clock. One is to follow a circuit such as those presented thus far with either a Schmitt trigger or a comparator operated as a zero-crossing detector. The *Schmitt trigger* is a circuit that produces a HIGH output when the input voltage exceeds a certain threshold and

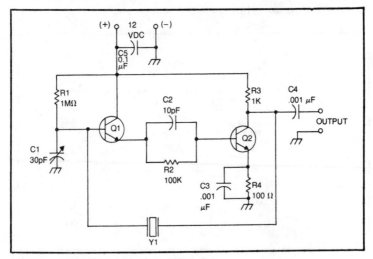

Fig. 31-10. Parallel-mode oscillator (aperiodic).

332

stays HIGH until the input voltage drops below another threshold. Very rarely are positive-going and negative-going thresholds the same, so there is always some hysteresis between the two values. Figure 31-11A shows the circuit symbol for a 7414 TTL Schmitt trigger. This IC device contains six Schmitt trigger sections that may be operated independently of each other. In the case of the TTL Schmitt trigger (see Fig. 31-11B), the output voltage must be compatible with TTL levels, so it must current sink at 0 volts and rise to +5 volts when a HIGH is demanded. The input impedance is around 6000 ohms. The thresholds for this device are different for positive-going and negative-going signals. When the signal is positive-going, the output of the Schmitt trigger will snap HIGH at 1.7 volts (point "A" in Fig. 31-11B). The output remains at the +5V level (HIGH) until the input signal passes the 0.9V level (point "B") on its way back down, or on the negative-going side of the waveform. Note that we said, "negative-going," not "negative." The trip point is a positive potential, but the derivative of the waveform at this point must be negative. The output of the Schmitt trigger is a train of nice, clean, sharp square waves that are compatible with TTL digital circuits.

We can simulate the action of the Schmitt trigger using the 4049 or 4050 devices in the CMOS line. This line of digital ICs works with power-supply levels up to ±15 volts, and output transitions occur only when the input signal passes the ((V+) − (V−))/2 point. If we select 0 volts and +12 volts as the two potentials, the output transitions occur when the input signal passes the (+12) − (0))/2, or +6V, point. The 4049 and 4050 devices are similar to each other, except that one is an inverter, and they are different from the rest of the CMOS line because their outputs are TTL-compatible. These, then, can also be used to square up the oscillator signal at frequencies to 5 MHz.

The comparator idea is shown in Fig. 31-12. Indeed, an operational amplifier can be used as a comparator—and frequently they are—if the feedback resistor is removed. With no feedback resistor, the gain of the amplifier is essentially the open-loop gain of the op amp. This value could be as low as 20,000 or as high as several million. Once the *difference* between the two input terminal voltages is more than a few millivolts, the output will snap HIGH in either the positive or negative direction, depending upon the polarity of the differential input voltage. Figure 31-12 shows the use of a special comparator IC (the LM-311) that is frequently seen. The LM-311 has an open-collector output stage, so a pull-up

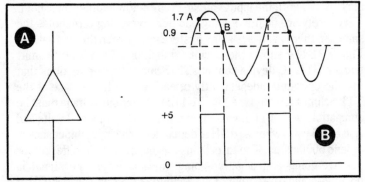

Fig. 31-11. Schmitt trigger symbol (A) and timing diagram (B).

resistor (R1) of 2200 ohms is needed to make it compatible with +5V TTL circuits. The LM-311 can be operated with any combination of power supplies up to ± 15 volts, but here we opt for the monopolarity TTL scheme in which +5 volts is applied to pin No. 8 and both pins 1 and 4 are grounded.

To operate the comparator as a zero-crossing detector, connect the noninverting (+) input to ground. When the input signal from the sine wave oscillator is zero, the output of the LM-311 will also be zero. But when the sine wave is nonzero the output of the LM-311 snaps HIGH to +5 volts. The result is an output of pulses that can be used as the clock in a digital circuit.

Another approach is to use an actual TTL oscillator. One of the most popular circuit configurations is shown in Fig. 31-13.

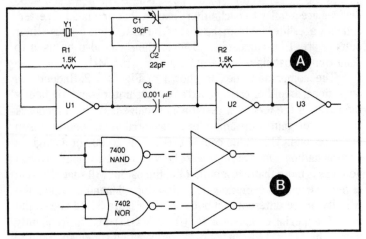

Fig. 31-12. Schmitt trigger circuit.

334

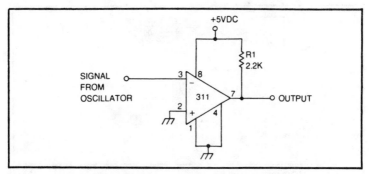

Fig. 31-13. TTL crystal oscillator using gates or inverter sections.

Here we cross-connect two TTL inverter stages, U1 and U2, with a crystal Y1. A resistor in each stage biases the normally digital gates into a region where they operate as amplifiers. Inverter stage U3 is used as a buffer and also serves to clean up the output signal to make it look more like something that a TTL digital circuit wants to see.

In most cases, we would not use the inverters shown in Fig. 31-13A, but either a NAND gate or a NOR gate. These gates are in almost all digital circuits and can be made to act as inverters. The NAND gate produces a HIGH output if either input is LOW, and will produce a LOW output *only* when *both* inputs are HIGH. The NOR gate produces a LOW output if either input is HIGH, and a HIGH output *only* if *both* inputs are LOW. We can make a NAND gate act as an inverter by connecting the two inputs together. That way, they both go HIGH and LOW together. When the inputs are LOW, the output will be HIGH, or inverted. But when the inputs are HIGH together, the output will be LOW, or inverted. The NOR gate can be made to act as an inverter by grounding one input or by

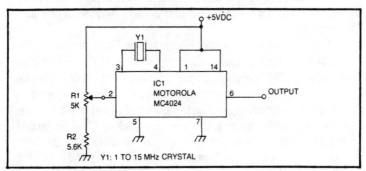

Fig. 31-14. TTL crystal oscillator using the Motorola MC4024P device.

335

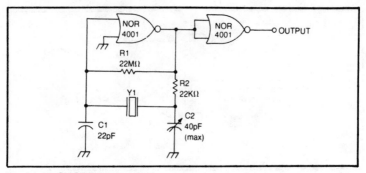

Fig. 31-15. CMOS crystal oscillator.

tying the two inputs together. The reason why it works follows from the properties of the NOR gate.

Several TTL-family IC devices are intended to operate as oscillators. The circuit in Fig. 31-14 shows the use of a Motorola MC4024 IC. This chip is intended to operate as a voltage-controlled oscillator (VCO) in a phase-locked loop (PLL), but it can be pressed into service as a crystal oscillator. Normally, in VCO operation, a capacitor is connected between pin Nos. 3 and 4 and a control voltage is applied to pin No. 2. The output frequency is variable over a 3:1 range by changing the control voltage on pin No. 2, while large changes in frequency may require a new value of capacitor. In service as a crystal oscillator, however, a crystal is connected across pin Nos. 3 and 4. In some texts, the control voltage pin is strapped to the +5V line, but it has been this author's experience that such an arrangement makes it hard to start the oscillator in some cases. The circuit shown in Fig. 31-14 is a little better in this respect because it allows us to adjust the pin No. 2 voltage to a value that will make the crystal start easier. The output is taken from pin No. 6 and is a chain of TTL-compatible square waves. This IC is a reliable chip that is easy to tame in practical situations.

Figure 31-15 shows a CMOS crystal oscillator circuit for the 0.5-MHz to 2.0-MHz region. This circuit is from the RCA Corp. (inventors of CMOS) applications literature on their CD4000-series of CMOS digital integrated circuits. This circuit is based on the CD4001 inverter stage. The frequency of the oscillator can be adjusted to a precise value with trimmer capacitor C2. The second NOR gate serves as an output buffer. If the circuit is operated from a single +5V power supply (it can take up to 15 volts), the oscillator is TTL-compatible and will drive up to one TTL device.

Chapter 32
Solid-State
Microwave Devices

Engineering advances of the past four decades have dramatically increased the amount of radio spectrum that is available for practical exploitation. In the era just after World War I, commercial and military interests considered frequencies with wavelengths less than 200 meters useless. In fact, until the later 1930s many publications referred to all frequencies with wavelengths shorter than 10 meters as *ultrahigh frequency* (UHF), a designation that we use today for frequencies in the 300MHz to 100-MHz region. Even today, a holdover of the older meaning of the rubic "UHF" is seen on German-made FM broadcast radio receivers which mark the FM band (88MHz to 108MHz) with *U* or *UKW* after the German words for ultrahigh frequency, *ultra kurz welle*. Several design problems had to be solved before the UHF and microwave spectrum could be properly exploited outside the laboratory. Today, some engineers—perhaps mindful of the increasing frequency of operation of microwave devices—have referred to upper gigahertz frequency as UHF: *unbelievably high frequencies*!

Before discussing the matter of solid-state microwave devices, let us first examine the history of microwave research, with particular emphasis on the device problems that had to be solved first. The generation of ultrahigh frequencies is not, as some might believe, a strictly modern phenomenon. Heinrich Hertz conducted many of his early (1887) experiments using frequencies in the 60-MHz to 500MHz range. Marconi, usually considered the father of commercial wireless telegraphy, used Hertzian apparatus on frequencies to 500 MHz but soon switched to very low frequencies (VLF), those with wavelengths longer than 200 meters, when he discovered that they were more effective with the primitive radio equipment than available.

By 1930, laboratory workers had succeeded in generating frequencies as high as 75 gigahertz, but not in a commercially important manner. The power available at those frequencies was minuscule because it represented the highest detectable harmonics of spark-gap rf power generators that operated at a considerably lower fundamental frequency.

The generation of large amounts of rf power in the decades preceding World War II required one of three available technologies: *Spark-gaps, Alexanderson alternators* and *vacuum tubes*. Of these, only the spark-gap transmitter was capable of generating microwave power, and this only at the expense of huge amounts of energy at subharmonically related frequencies below the microwave region. It would not be unreasonable to assume that *microwatts* of energy in the low microwave region required *kilowatts* of energy from a 1-MHz spark-gap.

The Alexanderson alternator was a mechanical electrical alternator that produced frequencies in the VLF region. The frequency was determined by the number of poles and the speed of rotation. The alternator could be keyed for radiotelegraphy by switching the current to the field winding. In 1916, Navy engineers working at radio station NAA (then located in Arlington, Virginia) succeeded in amplitude modulating the output of an Alexanderson alternator. Although the alternator was used to generate large amounts of rf power, it was limited in frequency by mechanical problems.

Vacuum tubes of the post-World War I era would ordinarily not operate at UHF or microwave frequencies. They did, however, seem to represent the best approach to researchers hoping for UHF/microwave operation. There were two major problems with vacuum tubes that kept operating frequency low: *interelectrode capacitance* and *electron transit time*.

There are two approaches to reducing interelectrode capacitance. One is the electrode size and geometry, while the other is the interelectrode spacing. Reducing the electrode size is not always viable, however, because this approach also limits the output power; smaller electrodes are not competent to dissipate the heat generated by impacting electrons.

Increasing the element spacing in a vacuum tube will decrease the capacitance, but it will also increase the transit time. Geometry, size and spacing factors could account for some of the progress in vacuum-tube design, such that operating frequencies to 450 MHz were used in World War II.

As early as the 1920s investigators had noted that electron transit time seemed to be a fundamental limitation in the design of microwave vacuum tubes. Electron transit time is the time required for an electron to travel from the cathode to the anode. Proper grid control of the electron stream requires that the period of the alternating voltage applied to the grid be short compared with the electron transit time. It proved relatively easy to design around the transit time problem at frequencies to 200 MHz, with 750 MHz reportedly attained in the laboratory by Boyd and Phelps (1927). But operation at frequencies greater than 200 MHz proved difficult to achieve on a commercial scale.

A solution to the transit time problem was proposed in 1920 by Barkhausen and Kurz of Germany. The *Barkhausen-Kurz oscillator* (BHO) used a special configuration vacuum tube to generate 700-MHz signals. In the BKO tube (Fig. 32-1), the grid is made postitive with respect to the cathode *and the anode*, exactly the opposite of the usual arrangement. Electrons emitted by the thermionic cathode are attracted by the positive potential on the control grid. Some of these electrons will strike the grid, but most are accelerated through the grid structure toward the anode. Shortly after passing the grid, however, they are repelled by the negative potential on the anode, and are deflected back toward the grid. The electrons will oscillate in an elliptical path around the grid structure. Output power can be obtained by connecting the grid to a load and will consist of the minority of electrons in the grid path that actually strike the grid.

American development of the BKO was hampered by an import embargo imposed upon nonAmerican vacuum tubes during the famous radio patent wars of the 1920s. The BKO could be made from an ordinary vacuum tube, but it required a cylindrical coaxial anode-grid configuration, as used in German vacuum tubes. American tubes, however, used a flat, sandwich-like structure that was not amenable to BKO operation.

European work with the BKO, however, continued. Kohl, in Germany, succeeded in generating a 6-GHz signal in 1928. Esau used BKO devices to make full-duplex communications across the

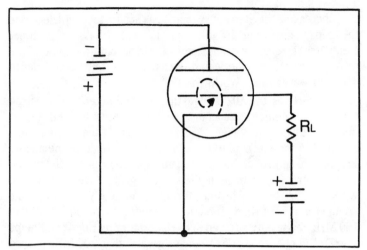

Fig. 32-1. BKO operation of a vacuum tube.

English channel in 1930. The frequency used in these tests was 1.6 GHz.

The BKO was one of the first devices to overcome the transit time limitation by making it work to advantage. The BKO overcame the time problem by keeping the electrons oscillating in a circular path in an electric field. The vacuum-tube grid, however, was a factor that limited available output power; its small dimensions often resulted in the grid running white hot during BKO operation.

A solution to this problem was proposed in 1921 by A. W. Hull: Delete the grid altogether and keep the electrons in orbit using a *magnetic* field. Hull's original *magnetron* has been much modified over the years, but the basic principle is still used today. Most microwave ovens, for example, use magnetron oscillators. One of the first modifications to the Hull magnetron was made by Yagi and Okabe of Japan. The Yagi-Okabe magentron achieved greater output power and higher operating frequencies by splitting the Hull anode into two or more sections. There were, however, still considerable design problems to solve because the small dimensions of the components required for microwave operation made the output power somewhat limited. By the mid-1930s, however, Cleeton and Williams, working at the University of Michigan, achieved operation at 50 MHz. Over the years, power and frequency have increased, making the magnetron one of the primary sources of microwave energy.

The power-versus-frequency dilemma seemed unsolvable for several years. But in the mid-1930s, several investigators simultaneously reached similar solutions. Dr. W. W. Hansen of Stanford University and Drs. A. and O. Heil began to think in terms of tuning the transit time to advantage through the mechanism of *velocity modulation* of the electron beam. The Heils proposed in 1935 to use the electron transit time to control the electron stream. The heating problem was not solved but was *avoided* because the electrons would not actually strike the control electrodes.

Russell and Sigurd Varian extended Dr. Hansen's work into the practical world in 1937 when they used Hansen's calculations to build the first *klystron* vacuum tube. This device used the transit time and the deceleration of bunched electrons to generate microwave rf energy. Velocity modulation of the electron stream in the klystron produces the bunching effect. It is the time between the arrival of successive bunches at a collector anode that determines the operating frequency of the klystron. Arrival of each bunch represents one cycle of rf energy.

The background material presented thus far is intended to demonstrate some of the problems that required solution before microwave frequencies could be properly exploited. The development of semiconductor devices saw similar, if not identical, problems. The high-frequency response of bipolar transistors, for example, was limited by the transit time of charge carriers (electrons or holes) across the base region. Attempts at reducing the width of the base region in order to decrease transit time produced additional problems, such as increased capacitance and decreased tolerance to reverse bias potentials.

But even thin base regions could not solve the problem. In semiconductor materials a property called *electron saturation velocity* is analogous to a similar problem in vacuum tubes. This limitation seems to be a fundamental limit to the high-frequency operation of bipolar transistors. But like the vacuum tube transit time problem, this problem can be turned into an advantage and used to create microwave oscillations.

TRANSFERRED ELECTRON DEVICES

John Gunn of *International Business Machines Corp.* was studying the propertries of n-type gallium arsenside (GaAs) material in 1963. He noticed that the current through the material would become unstable if the applied voltage were increased above a certain threshold potential. It was discovered that the current would pulsate at microwave frequencies if the E-field were above this point.

Gunn suspected that a negative resistance phenomenon was responsible for the observed observation. Negative resistance generators can be made to oscillate under the right circumstances. Gunn speculated that the negative resistance phenomenon was due to a loss of electron mobility at the higher applied voltage. This theory can be inferred from the fact that some materials, such as GaAs, permit electrons to exist in either of two, rather than just one, conduction bands (Fig. 32-2). In the lowest conduction band, the electron effective mass and electron energy are low. The energy level is close to the minimum allowed for conduction bands in that material. Electron mobility in this conduction band is high—on the order of 8000 cm^2/V-s—so the material will act like an ordinary ohmic resistance.

If the electric field is increased to approximately 3 to 3.5 kV/cm, the electrons will become more energetic and will transfer to the higher conduction band. An energy level of 0.35 eV separates the lower of the two conduction bands. The electron effective mass increases in the higher conduction band, while mobility and drift velocity are decreased.

Figure 32-3 shows that an increasing number of electrons are scattered into the low-mobility conduction band as the applied potential increases above a certain threshold voltage, V_{th}. At potentials less than the threshold potential, the electron velocity increases linearly with the applied voltage. This behavior is exactly as expected in any material that obeys Ohm's law. But at potentials greater than V_{th}, the number of electrons transferred to the low-mobility—low-velocity conduction band increases with increasing voltage. This phenomenon causes the net electron velocity to drop, creating a negative resistance region between V_{th} and a saturation potential, V_s. Microwave oscillators that depend upon the transfer of electrons between high-mobility and low-mobility conduction bands are called *transferred electron oscillators* (TEO).

The description of negative resistance operation adequately justifies the claim that the GaAs material will oscillate, but it does not explain the structure of the Gunn device or the mechanism of oscillations in these devices. There are two different *modes* of oscillation in the Gunn device.

Incidentally, it has become common practice to call the Gunn device a *Gunn diode*, but this is not strictly correct: The Gunn

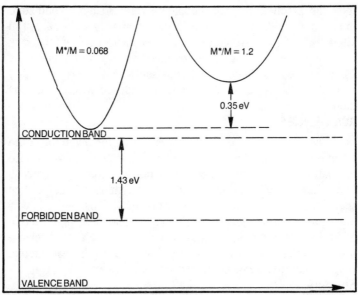

Fig. 32-2. Conduction bands for GaAs used as a transferred electron oscillator.

device is *not* a diode. In the Gunn device, oscillations are used in a bulk medium, not a pn junction. The diode-like structure of the Gunn device shown in Fig. 32-4 is not an ordinary pn junction. The end sections are not active, but are intended to facilitate ohmic contact between the electrodes and the active center region. Also,

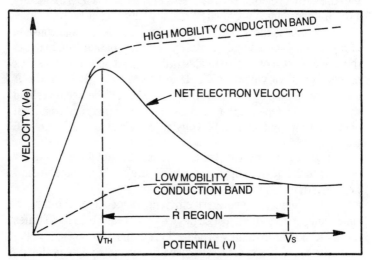

Fig. 32-3. Conduction as potential increases shows a negative resistance.

it is the practice of some people to refer to most solid-state microwave oscillators as Gunn diodes, when, in fact, they are of an altogether different structure. One recent advertisement listed as a "Gunnplexer" a device that contained an IMPATT diode. The Gunn device is a diode only in the sense that it has two (di−) electrodes (ode), but it should not be confused with pn junction diodes.

GUNN DEVICES

The observations of Gunn in 1963 led to the invention of the Gunn device as a microwave oscillator. But negative resistance oscillators were the subject of speculation by Shockley as early as 1954. Esaki (1957) had produced a two-terminal "diode" device that exhibited a negative resistance property. This device is now known as the *tunnel diode* or *Esaki diode*. Sommers suggested in 1959 that the Esaki diode would have microwave applications; a prediction that has proven accurate. It was against this background that the Gunn device was invented in the mid-1960s.

Structure

The basic structure of the Gunn device is shown in Fig. 32-4. This device consists of three basic sections, labeled *A, B,* and *C*. Region-*A* provides ohmic contact with the active center region, so it should be made of relatively low resistivity (0.001 ohm-cm) material. Its purpose is to ensure good contact with the end electrode and to prevent metallic ion migration from the electrode into the active region. The thickness of region-A is approximately 1 to 2 microns, and it is grown epitaxially onto region-B.

Region-B is the active region in the Gunn device and consists of n-type gallium arsenide material. The oscillating frequency of the device in one of its two modes is dependent upon the thickness of this region, varying from 6 GHz to 18 μM to 18 gHz at 6 μM. A frequently quoted figure is 10 GHz at 10 μM. The device threshold voltage is also dependent upon the thickness of this region and will vary from approximately 1.95 volts at 18 GHz up to 5.85 volts at 6 Ghz.

Region-C is the substrate layer and is metallized to allow bonding to the device support structure, the diode package. Again, low-resistivity GaAs material is used.

Gunn devices are not very efficient, especially in the transit-time mode of operation. These devices may require 20 to 50 times more DC power than they produce in rf output. As a result of the low efficiency, the substrate is usually bonded to a heat sinking package.

Modes of Oscillation

The Gunn device can operate in either of two different modes: *transit-time* (or *Gunn*) *mode* and the *limited space-charge* (or *delayed transit-time*) *mode*. The transit-time mode depends upon the thickness of the active region for operating frequency, but it does not require an external tank circuit for proper operation. The delayed-transit time mode requires an external tank circuit, such as a tuned cavity, but it is frequency-flexible and operates with more efficiency.

Transit-Time (Gunn) Mode. When a Gunn device is biased below the threshold potential, V_{th}, the electric field will be uniform throughout the device (Fig. 32-5). The Gunn device will operate as an ordinary positive resistance in the region; the current will increase proportionally with the increasing voltage.

But consider the situation (Fig. 32-6) when the Gunn device is biased to the threshold potential, V_{th}. Under this condition, electrons are injected into the cathode end of the material faster than they are collected at the anode end. This causes a domain to build up that is rich in excess electrons on one side and deficient in electrons at the other side (the depleted region is on the anode side of the domain). This domain drifts through the length of region-A in the Gunn device until it is collected at the anode end. A new domain forms as the old domain is collected at the anode.

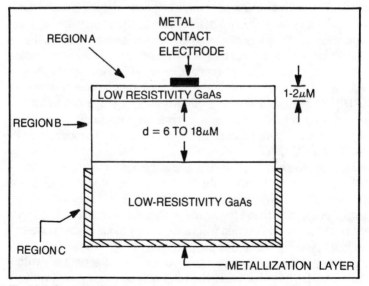

Fig. 32-4. Gunn "diode."

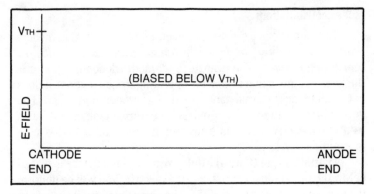

Fig. 32-5. Gunn device below V^{th}

The current output from the Gunn device maintains a low background level (Fig. 32-7) until the domain is collected at the anode. At that instant a brief current pulse is generated in the output circuit.

The period of time between output current pulse peaks is the drift time for that particular sample of material. The period—hence, the operating frequency—is dependent upon the path length and the drift velocity (on the order of 10^7 cm/s) of the domain.

Delayed-Domain Mode. The delayed-domain, or limited space-charge (LSA), mode is more efficient than the transit-time mode. The transit-time mode, while elegantly simple, suffers from low efficiency and a frequency limitation that is determined by the thickness of the active region. The delayed-domain mode allows the Gunn device to adapt to the frequency of an external tank circuit, such as a high-Q resonant cavity. Figure 32-8 shows an equivalent circuit in which the negative differential resistance (NDR) of the Gunn device is shown as a negative conductance placed in parallel with the LC tank circuit. Conductance G_o represents the ohmic losses in the tank circuit. The circuit will oscillate if $-G << G_o$.

Suppose that we bias the Gunn device at some potential greater than the threshold voltage. The domain-creation phenomenon of the transit-time mode will cause several initial output pulses that serve to excite, or ring, the external tank circuit into oscillation. This action will cause a constinous rf sine wave to build up (see Fig. 32-9) that has a frequency equal to the resonant frequency of the tank circuit. The rf voltage adds algebraically with the static DC bias such that the total bias is greater on positive peaks and less on negative peaks. The value of the static DC bias

346

must be carefully adjusted so that the total bias drops below V_{th} on negative peaks of the rf cycle, yet will remain above the minimum sustaining potential. Whenever the total bias (the sum of DC and rf voltages) is less than the threshold potential, the domains are quenched. If the previous domain reaches the anode while the bias is below V_{th}, the creation of the next domain is *delayed* until the rf cycle brings the bias back above the threshold potential. This phenomenon causes the output current pulse period to adjust automatically to the period of the external tank circuit. Figure 32-10 shows LSA output pulses. We can use the frequency agility of the delayed-domain mode to frequency modulate the device, or make it subject to *automatic frequency control* (afc) operation, by manipulation of the DC bias potential.

The delayed-domain, or LSA, mode is considerably more efficient than the transit-time mode. The output power available in the transit-time mode is usually less than 1000 milliwatts, with efficiencies on the order of only 1 to 5 percent. The delayed-domain mode, on the other hand, can deliver peak powers up to several hundred watts (duty cycle of 0.01 or less). The operating frequency

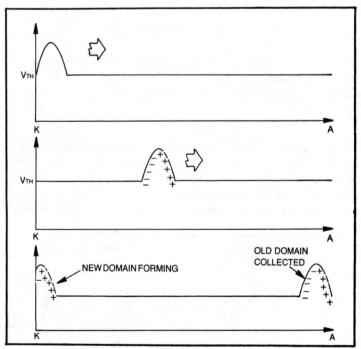

Fig. 32-6. Creation of domains.

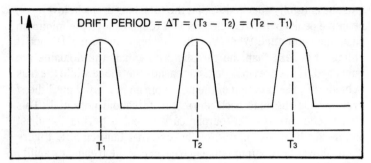

Fig. 32-7. Output pulses.

of the transit-time mode is determined by the length of the active region and the saturated velocity of the electrons ($1/f = T = L/V_{sat}$), so it is not variable. The delayed-domain mode, however, will adjust itself to the resonant frequency of the high-Q tank circuit in which it is operated. We can often adjust the operating frequency of one octave by adjusting the tank dimensions. The frequency of operation in the delayed-domain mode can be found from:

$$T = \frac{1}{f}$$
$$= 2\pi\sqrt{LC} + \frac{L}{R_0 (V_v/V_{th})}$$

where T is the period, f is the frequency, L and C are the cavity resonance parameters, R_0 is the low-field resistance, V_b is the DC bias potential, and V_{th} is the threshold potential.

Gunn Oscillators

The Gunn device will oscillate in the transit-time mode using only a simple resistance for the load. The efficiency in this mode, however, is only 1 to 5 percent, so relatively large amounts of DC power are required to generate small amounts of rf power. If we place the Gunn device inside a resonant cavity and bias the device

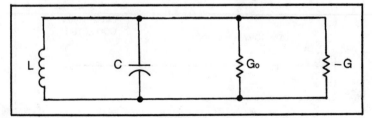

Fig. 32-8. Tank circuit in parallel with negative conductance.

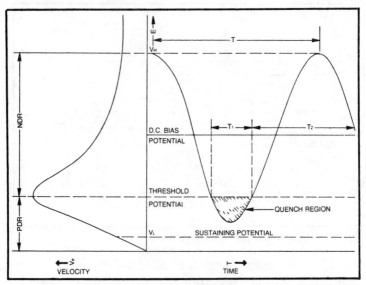

Fig. 32-9. Gunn device delayed domain or LSA mode.

for the delayed-domain mode, we will obtain better efficiency and some flexibility of the operating frequency.

Figures 32-11 and 32-12 show two methods for mounting a Gunn device inside a resonant cavity. Figure 32-11 shows a cutaway view of a coaxial cavity. The length of the cavity is one-half wavelength, while the base of the Gunn device is placed at the

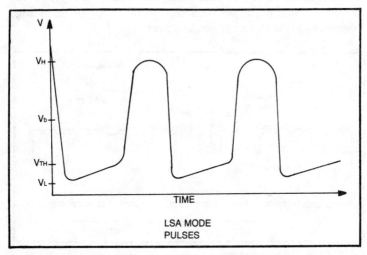

Fig. 32-10. LSA output pulses.

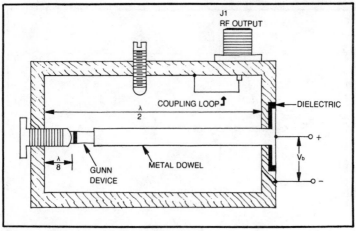

Fig. 32-11. Gunn device in waveguide section.

one-eighth wavelength point. A conductive dowel supports the Gunn device and connects it to the ends of the cavity; the dowel is also the center conductor of the coaxial cavity.

A tuning screw is used to vary the operating frequency of the device. It effectively changes the dimensions of the cavity and is capable of fine tuning the operating frequency over a small range.

The oscillations on the inside of the cavity are coupled to the outside world through a short coupling loop that is situated parallel to the dowel center conductor. The load impedance of the Gunn device is set by the position of the coupling loop and is adjusted for the best compromise between the stability of the operating frequency and the maximum output power.

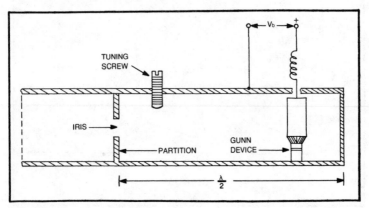

Fig. 32-12. Gunn device in iris waveguide.

350

The coaxial cavity, while simple, suffers from a few basic problems. It is a low-Q tank and is more sensitive to factors such as temperature and load impedance variations. The Gunn device in a coaxial cavity may also tend to oscillate on a harmonic of the tank frequency.

A rectangular waveguide (Fig. 32-12) can also be used as a tuned cavity if one end is blocked off and the Gunn device is placed at the one-eighth wavelength point. DC bias is provided to the Gunn device through an rf choke that is designed for microwave operation.

The dimensions of the cavity are determined by the placement of a partition. Energy from the cavity is coupled into the waveguide transmission line through an opening called an *iris*. The size of this iris is a trade-off between maximum output power and a sensitivity to changes in the load and internal impedances of the Gunn device.

IMPATT DEVICES

The IMPATT (impact avalanche transit time) diode was proposed in 1953 by W. T. Read of *Bell Laboratories*. Read's suggestion was that the phase delay in a pn junction diode between an applied rf voltage and an avalanching current could be used for negative resistance operation at microwave frequencies. Read's model diode would have carriers drifting through a depletion region to cause the negative resistance. Fabrication difficulties

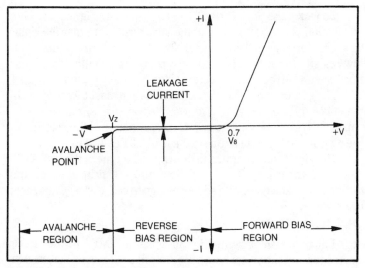

Fig. 32-13. Avalanche diode characteristic.

prevented the construction of a working Read diode until the mid-1960s. In 1965, however, R.L. Johnson of Bell Labs verified the validity of Read's model when he generated approximately 80 milliwatts of rf energy at 12 GHz from a silicon pn junction diode. Read's diode depends upon impact avalanche and transit time phenomena, so it was given the acronym, IMPATT. It has now been recognized that Read's structure is just one of several that will result in IMPATT operation.

Avalanche Phenomena

Figure 32-13 shows the I-versus-V curve for a pn junction diode. For our present purposes, we will consider only operation in the reverse bias region, the region in which V is less than zero. There is a critical breakdown voltage, V_z, in the reverse bias region. At reverse potentials less than this value, the current through the pn junction is very small; it is a tiny leakage current, in fact. But the current suddenly increases when the voltage exceeds V_z: The junction is operating in avalanche. The increased current is due to secondary emission or avalanche multiplication, in which electrons of the leadage current have a high probability of being multiplied. The result is a very rapid increase in reverse current. In ordinary signal or rectifier diodes, the avalanche phenomenon can be destructive. Certain types of diodes, however, are able to control the avalanche process by using properly doped semiconductor material. Zener diodes and controlled avalanche rectifiers are in this category.

Consider the pnn⁺ IMPATT diode shown in Fig. 32-14. The pn junction is on the left side of the structure. Note that the right hand junction is an n-n⁺ structure. The n⁺ region forms a contact of low resistivity for the electrode and prevents metallic ion migration (much as in the Gunn structure) into the active region.

The center n region is the active zone and must be doped such that it is fully depleted at breakdown. We want to insure that a very small electrical field will cause velocity saturation of the electrons. We may define the transit time of the IMPATT as $T = L/V_{sat}$.

The electrons generated in the avalanche zone of the IMPATT diode shown in Fig. 32-14 will flow into the drift zone of the n-region. It takes very little added voltage to cause a large increase in current in this mode.

IMPATT Oscillation

Let us consider a situation in which an IMPATT device is biased to a potential just below V_z; i.e., in the reverse bias region but not quite to the avalanche point. We must select a bias such that

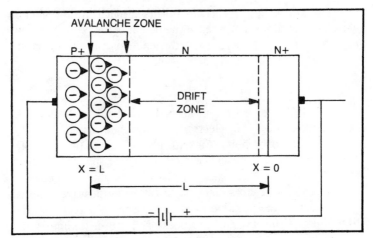

Fig. 32-14. Avalanche phenomenon for IMPATT diode.

a small added potential will throw the device into the avalanche region. Let us further assume that the IMPATT device is operated in parallel with a high-Q resonant tank circuit, such as when the IMPATT device is operated inside a resonant cavity. The reverse biased pn junction will create a noise signal that shock excites the tank circuit into oscillation. The rf voltage produced by the oscillatory tank is added to the bias voltage, causing the diode to go into the avalanche mode on positive peaks of the cycle. The junction will exhibit negative resistance operation if the rf current can be made to lag the rf current by 90 degrees or more.

The number of electrons generated by avalanche multiplication is a function of the applied voltage (Fig. 32-15A) and the number of charge carriers present. Because of this dual dependence, the avalanche current pulse (Fig. 32-15B) continues to increase even after the rf voltage cycle has passed its peak. During this process, the charge density at the avalanche point grows exponentially while the avalanche charge current (Fig. 32-15C) drifts toward the other end of the drift zone at velocity $V_s = 10^7$ cm/s.

Does the IMPATT produce negative resistance? Note that the current reaches a peak (Fig. 32-15C) as the sine wave rf voltage goes through its zero crossing (Fig. 32-15A), which is a 90-degree delay with respect to the voltage peak. The criterion for negative resistance is a phase difference of 90 degrees or more between the applied voltage and the series current, so we may conclude that the IMPATT is a negative resistance device.

The avalanche current pulse drifts through the n-region at velocity V_s until it is collected at the output of the N^+ region. The drift time, then, is proportional to the length of the drift region and inversely proportional to the saturation velocity:

$$T = \frac{L}{V_s}$$

where T is the drift time in seconds, L is the path length in centimeters, and V_s is the electron saturation velocity (10^7 cm/s.

The pulse of current in the external tank circuit (Fig. 32-15D) is semisquare and represents a current lag over applied voltage of more than 90 degrees. These two factors are shown together in Fig. 32-16. Two factors combine to cause the positive external current during the negative excursions of the rf waveform: The time delay of the avalanche process and the drift time of the avalanche charge. Instead of absorbing energy, in the manner of a

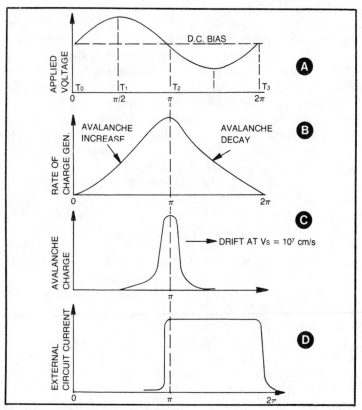

Fig. 32-15. Timing diagrams for IMPATT.

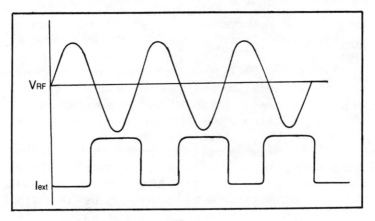

Fig. 32-16. Output pulses from IMPATT.

positive, or ohmic, resistance, the IMPATT offers a negative resistance.

The IMPATT device, which was shown previously, is known as a *single-drift*. But an avalanching pn junction produces both kinds of charge carriers: holes and electrons. The single-drift IMPATT uses only the electrons and returns the holes to the cathode p-region. This fact limits the efficiency of the single-drift devices to less than 15 percent.

Double-Drift IMPATTs

Greater efficiency is obtained through the use of a double-drift IMPATT device, such as shown in Fig. 32-17. This structure is a $p^+-p-n-n^+$ in which the avalanche region brackets the pn junction. The p^+ zone serves as an ohmic contact for hole charge carriers, while the n^+ region serves the same purpose for electrons. The output efficiency is increased over that of the single-drift variety because the holes drift across the p-zone very nearly in phase with the electrons drifting across the n-zone.

IMPATT Applications

The previous discussion has demonstrated that the IMPATT device will function as an oscillator at microwave frequencies. If an IMPATT is placed inside a high-Q resonant cavity and biased with a DC potential slightly below the avalanche potential, noise pulses will ring the cavity to produce the rf sine wave that actually drives the junction into the IMPATT mode of oscillation. IMPATT operation occurs because the voltage of the ringing waveform (an rf signal) algebraically adds with the static DC bias, causing the junction to go into avalanche mode on peaks of the rf cycle. If the device is correctly biased, then, the junction will be in the avalanche condition for most of the positive half of the rf sinewave excursion.

Although the IMPATT device is an oscillator that is capable of producing substantial peak pulse powers at microwave frequencies, it is not universally applied because it is a noisy source; avalanching is a noisy process. For this reason, one does not ordinarily see IMPATTs as receiver local oscillators.

IMPATTs are used primarily at frequencies above 3 or 4 GHz, with frequencies up to 100 GHz having been obtained. Many high-power IMPATTs require operating potentials between 75 and 150 VDC; a fact seen as a disadvantage by some. Also, IMPATTs are usually operated from constant-current power supplied, also a potential disadvantage. The efficiencies obtained from IMPATTs range from 12 to 15 percent for single-drift devices, and 20 to 30 percent for double-drift devices made of GaAs material.

The applications of the IMPATT are not limited to oscillator service. There is one report of IMPATTs being used as a microwave frequency multiplier. Many IMPATTs are used as amplifiers; in fact, it has been claimed that most IMPATT

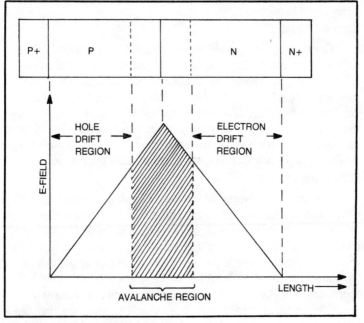

Fig. 32-17. Double-drift device (IMPATT).

applications are as amplifiers, not as oscillators. IMPATT amplifiers have only one port, so they must be coupled to a *circulator* to isolate input and output ports of the amplifier (see Fig. 32-18). This type of amplifier is called a *reflection amplifier.*

TRAPATT DIODES

IMPATT diodes are generally limited to operation at frequencies above 3 or 4 GHz. The problem of lower operating frequencies (0.9 GHz to 3 GHz) is one of finding a method for stretching the duration of the transit time. Until 1967, it was difficult to use solid-state devices to generate any significant amount of power in the 1-GHz region. In 1967, however, engineers working for *RCA* succeeded in exciting an IMPATT-like device into a different mode of operation. One set of trials produced pulse powers of 425 watts with an efficiency of 25 percent. Further work with this new mode yielded efficiencies up to 60 percent, with later work producing efficiencies as high as 75 percent. Tuned tank circuits developed at RCA in that era permitted a tuning range that was continous over 0.9 GHz to 1.5 GHz.

It appeared that the problem of increasing the transit time had been solved, but no one really knew why! At the time the basic work on the TRAPATT device was going on, there was no good theory that explained the observed behavior. Workers at *RCA* dubbed the new mode the *anomalous mode,* perhaps reflecting the fact that they had no theory of operation.

At least two different theories were advanced to explain the behavior of the anomalous mode. *Bell Labs* advanced the theory

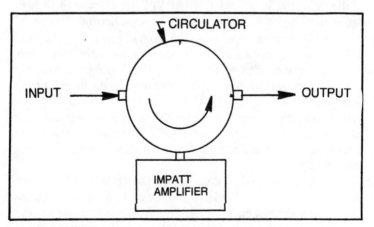

Fig. 32-18. IMPATT amplifier.

that the high efficiency and lowered frequency of operation was explained by the fact that a trapped plasma was created in the device between sweeps of the IMPATT mode of operation. The theory held that the trapped plasma shielded the charge carriers from the external voltage field, causing them to drift out of the plasma at low velocity. This theory led to the acronym by which the device is now known: TRAPATT (trapped plasma triggered transit).

Cornell University physicists offered a somewhat different explanation of the high efficiencies observed. The Cornell theory maintained that avalanche resonance pumping was the responsible mechanism, so Cornell advanced the acronym, ARP. The Bell Labs view seems to have prevailed. The difficulty in determining the proper theory of operation caused a two-year delay between the first observations of the TRAPATT mode and the explanation of how it worked. Part of the problem is that TRAPATT operation is not amenable to small-signal analysis, so the correct theory had to be worked out somewhat more laboriously than would have otherwise been possible.

It has been demonstrated that ordinary, silicon pn junction diodes can be made to oscillate in the TRAPATT mode. It is, however, rather tricky to adjust such circuits, so they do not find much application. Most commercial TRAPATT devices use the p^+-n-n^+ structure of the single-drift IMPATT. A typical TRAPATT device is shown in Fig. 32-19.

TRAPATT Oscillators

The structure of the TRAPATT device is very similar to that of the IMPATT. In fact, some TRAPATT devices will oscillate in either the TRAPATT or the IMPATT modes, depending upon the bias and other circuit conditions. Numerous TRAPATT oscillators actually start out in the IMPATT mode for a few nanoseconds after turn-on, and then convert to the TRAPATT mode when certain circuit conditions are satisfied. In order to make the device switch from the IMPATT to the TRAPATT mode it needs to be driven hard with a current pulse. Because the rise time of this pulse must be very short, it is usual to use the IMPATT mode to generate the pulses; it is very difficult to obtain the rise time needed with external circuitry.

What is a trapped plasma and how does it behave? A plasma exists whenever there is a large density of charge carriers (holes and electrons) present. If the electric field in that region is very low, the plasma is said to be *trapped:* It takes a long time to sweep

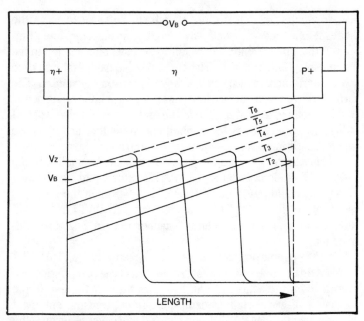

Fig. 32-19. TRAPATT device.

carriers out of the region under the influence of the electric field. The carrier velocity is considerably lower than the saturation velocity.

The device is biased to just punch-through. In other words, the depletion zone reaches through the entire length of the n-region but is biased at a point less than avalanche voltage V_z. The slope of the electric field is dependent upon the charge density. Because we are operating in the punch-through region, no free charge carriers will be present.

If we excite the TRAPATT diode with a large, fast rise time current pulse, I_o, a point in the constant bias field, V_B, will move as a avalanche shock front with a velocity, $V_x = -I_o/eN_D$. We find that the velocity of this shock front, V_x, can be faster than the saturated velocity of the holes and electrons, a phenomenon that is much like the behavior of water waves at the beach striking the shore with an angle other than 90 degrees. Typical times for the avalanche shock front to traverse the n-region are around 100 picoseconds.

Notice what happens to the terminal voltage in Fig. 32-19. Shortly after the initiation of each shock front, the terminal voltage drops from a very large value down to a very small value. The fall time of this drop is very rapid, so the TRAPATT operates as a very fast, low-impedance electronic switch. If we were to place the diode at one end of a half-wavelength transmission line (Fig. 32-20), this phenomenon will result in a pulse being applied to the transmission line at L = 0. The current pulse has a fast rise time and a nearly square waveshape, so it is rich in harmonics. The harmonics are taken out by a low-pass filter so that the fundamental could be applied to load Z_L. The value of the current pulse will be I $= V_B / Z_o$, where V_B is the applied bias potential, and Z_o is the characteristic impedance of the transmission line.

This description requires a current pulse to be applied to the TRAPATT diode before the TRAPATT mode could be realized. This pulse could easily be an IMPATT pulse when the device is operating in the IMPATT mode, but the TRAPATT mode will build up in a nearly exponential manner until it becomes self-sustaining.

The foregoing discussion does not explain how the TRAPATT mode could become self-oscillatory. For this type of operation, we must rely on the actions of the low-pass filter at the end of the resonant half-wavelength transmission. It will transmit energy at the fundamental TRAPATT frequency but will reflect energy at the harmonics of the fundamental frequency. These harmonics are reflected with phase reversal (reflection coefficient of −1 in the ideal case) so they will initiate another avalanche shock front. The small rise in the terminal voltage in Fig. 32-21A is caused by this returning reflection. The return pulse will, then, cause the TRAPATT mode oscillations to be continuous. In the typical TRAPATT oscillator, the device will begin in IMPATT operation. The IMPATT-mode oscillations will build up in an exponential

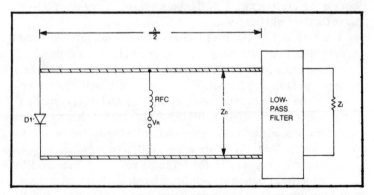

Fig. 32-20. If a diode is placed at one end of a half-wavelength transmission line, a pulse will be applied to the transmission line at L 0 .

manner until the current becomes large enough to trigger a shock front transit; hence, the use of triggered transit in the device acronym.

BARITT DEVICES

Consider the p^+-n-p^+ structure shown in Fig. 32-22A. This device is a pair of abrupt junctions back-to-back. One of these junctions will be slightly forward biased, while the other junction is slightly reverse biased. The flow of current under conditions when the bias voltage is less than the punch-through voltage will be limited by the ordinary leakage current of the reverse biased junction. If the bias is increased to the point where the device is operated in the punch-through mode, the depletion region exists across the entire n-region until it reaches the forward biased junction. This will cause all of the carriers (holes) at the forward biased junction to be swept across the n-region, causing the current to increase rapidly (Fig. 32-22B). This current can be used in a microwave oscillator provided that the field is large enough to

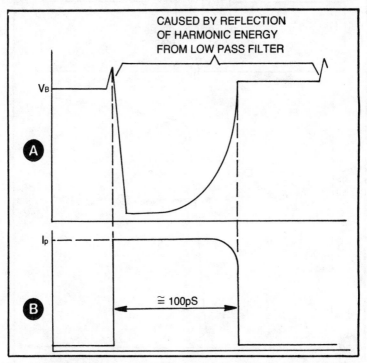

Fig. 32-21. The pulse that will be formed by the arrangement shown in Fig. 32-20. Voltage trace (A) and current trace (B) are shown.

make the holes drift across the n-region at the saturated velocity (10^7 cm/s), and the voltage applied to kept below the point at which avalanching will occur. Devices that use this phenomenon are called BARITT oscillators (barrier injection transit time).

Suppose that a BARITT device is biased with a DC potential close to the potential required for punch-through operation. Further suppose that this device is operated in parallel with a resonant tank, as was done in the discussion of the Gunn device earlier. Noise pulses will ring the tank circuit and cause an rf alternating potential to appear across the diode, which will add algebraically with the bias potential (Fig. 32-23A). When the total bias goes over the punch-through potential on positive excursions of the rf waveform, a sharp current pulse is injected (Fig. 32-23B). During the period when the injected current is peaking, the terminal current from the DC bias is added with it, causing a

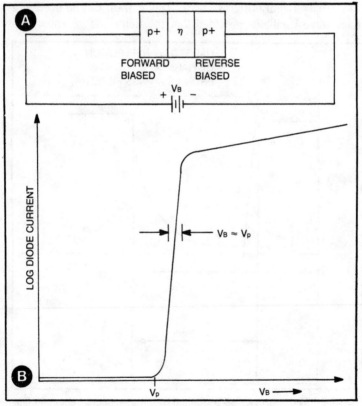

Fig. 32-22. BARITT device structure (A) and waveform (B).

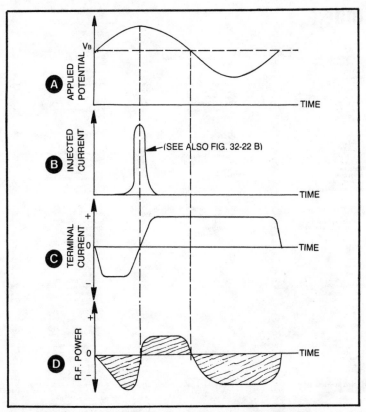

Fig. 32-23. Various BARITT waveforms: rf alternating potential across the diode (A), a sharp current pulse injection (B), current direction reversal when injected current is peaking (C) and power-versus-time waveform (D).

reversal of the current direction for that period (Fig. 32-23C). If we plot the power (product of Figs. 32-23A and 32-23C), we will obtain the power-versus-time waveform of Fig. 32-23D. Notice that for substantial periods during the cyclic excursion, the power is negative, meaning that the device is oscillating and will deliver energy to the external tank circuit. There is only a brief period in which the terminal current and terminal voltage are both positive $(T_2 - T_3)$, and this will limit the efficiency of the BARITT oscillator.

The BARITT device is a low-power microwave energy source. It is, in many ways, considered superior to the Gunn device for service such as microwave receiver local oscillators, doppler intrusion alarms and certain shoplifting protections devices.

Index

Edited by Raymond A. Collins